T0226159

Wireless Networks

Series Editor
Xuemin Sherman Shen, University of Waterloo, Waterloo, ON, Canada

The purpose of Springer's Wireless Networks book series is to establish the state of the art and set the course for future research and development in wireless communication networks. The scope of this series includes not only all aspects of wireless networks (including cellular networks, WiFi, sensor networks, and vehicular networks), but related areas such as cloud computing and big data. The series serves as a central source of references for wireless networks research and development. It aims to publish thorough and cohesive overviews on specific topics in wireless networks, as well as works that are larger in scope than survey articles and that contain more detailed background information. The series also provides coverage of advanced and timely topics worthy of monographs, contributed volumes, textbooks and handbooks.

** Indexing: Wireless Networks is indexed in EBSCO databases and DPLB **

Xi Jin • Changqing Xia • Chi Xu • Dong Li

Mixed-Criticality Industrial Wireless Networks

Springer

Xi Jin
Shenyang Institute of Automation, Chinese
Academy of Sciences
Shenyang, Liaoning, China

Changqing Xia
Shenyang Institute of Automation, Chinese
Academy of Sciences
Shenyang, Liaoning, China

Chi Xu
Shenyang Institute of Automation, Chinese
Academy of Sciences
Shenyang, Liaoning, China

Dong Li
Shenyang Institute of Automation, Chinese
Academy of Sciences
Shenyang, Liaoning, China

This work was supported by Chinese Academy of Sciences

ISSN 2366-1186   ISSN 2366-1445   (electronic)
Wireless Networks
ISBN 978-981-19-8924-7   ISBN 978-981-19-8922-3   (eBook)
https://doi.org/10.1007/978-981-19-8922-3

This Springer imprint is published by the registered company Springer Nature Singapore Pte Ltd.
The registered company address is: 152 Beach Road, #21-01/04 Gateway East, Singapore 189721, Singapore

Preface

Important tasks must be completed on time and with guaranteed quality; that is, the consensus reached by system designers and users. However, for too long, important tasks have often been given unnecessary urgency, and people intuitively believe that important tasks should be executed first so that their performance can be guaranteed. Actually, in most cases, their performance can be guaranteed even if they are executed later, and the "early" resources can be utilized for other, more urgent tasks. Therefore, confusing importance with urgency hinders the proper use of system resources. In 2007, mixed criticality was proposed to indicate that a system may contain tasks of various importance levels. Since then, system designers and users have distinguished between importance and urgency.

In the industrial field, due to the harsh environment they operate in, industrial wireless networks' quality of service (QoS) has always been a bottleneck restricting their applications. Therefore, this book introduces criticality to label important data, which is then allocated more transmission resources, ensuring that important data's QoS requirements can be met to the extent possible.

To help readers understand how to apply mixed criticality to industrial wireless networks, the content is divided into four parts. First, we introduce how to integrate the model of mixed-criticality data into industrial wireless networks (Chap. 1). Second, we explain how to analyze the schedulability of mixed-criticality data under existing scheduling algorithms (Chaps. 2 and 3). Third, we present a range of novel scheduling algorithms for mixed-criticality data (Chaps. 4, 5, and 6). Finally, we conclude this book and discuss future research directions (Chap. 7).

We hope that this book will inspire further research on mixed-criticality industrial wireless networks.

Shenyang, China
September 2022

Xi Jin
Changqing Xia
Chi Xu
Dong Li

Acknowledgments

We would like to thank Dr. Meng Zheng, Dr. Chunhe Song, Dr. Peng Zeng, Dr. Haibin Yu, Dr. Keyan Cao, Dr. Guoqi Xie, and Dr. Qingxu Deng for their helpful discussions and contributions to the research presented in this book, without which this book could not be completed.

This book was also supported in part by the National Natural Science Foundation of China under Grant 62133014, Grant 61972389, Grant 62022088, Grant 61903356, Grant 62173322, Grant 92067205, and Grant U1908212; in part by the Youth Innovation Promotion Association of the Chinese Academy of Sciences under Grant Y2021062, Grant 2019202, and Grant 2020207; in part by the National Key Research and Development Program of China under Grant 2020YFB1710900; in part by the International Partnership Program of Chinese Academy of Sciences under Grant 173321KYSB20200002; in part by the Central Guidance on Local Science and Technology Development Fund of Liaoning Province under Grant 2022JH6/100100013; and in part by the Liaoning Provincial Natural Science Foundation of China under Grant 2020-MS-034.

Contents

Chapter 1
Introduction

Abstract In this chapter, we first provide an overview of industrial wireless networks from the perspectives of industrial communication requirements and classical industrial wireless networks. Then, mixed criticality is introduced, and the role of mixed criticality in industrial wireless networks is also discussed. Last, we present the organization of this book.

1.1 Industrial Wireless Networks

The real economy with manufacturing as the core is a concrete manifestation of national economic strength and international competitiveness [1]. In the past 20 years, the manufacturing industry is undergoing the fourth industrial revolution, which is introducing intelligence into the industrial production process, commonly known as intelligent manufacturing (or smart manufacturing). Intelligent manufacturing has been widely recognized as an advanced technology to improve production efficiency and reduce production costs. To seize the commanding heights of the manufacturing industry, major industrial countries are striving to improve the intelligence of the manufacturing industry, and have issued relevant development initiatives, such as Germany's "Industry 4.0", China's "Made in China 2025", the United States' "Advanced Manufacturing Partnership", the United Kingdom's "Future of Manufacturing", and France's "Industrie du Futur" (as shown in Fig. 1.1). What these initiatives have in common is that they must be network-based.

Modern manufacturing has involved a large number of elements, such as machines, people, raw materials, etc. In order to make these elements form a whole system, networks must be used in industrial systems [2], called industrial networks. Industrial networks connect all elements together to implement the collection of industrial data and the sending of control commands. These uplink and downlink processes constitute a closed-loop control that enables intelligent manufacturing. Therefore, industrial networks are one of the important infrastructures of intelligent manufacturing.

X. Jin et al., *Mixed-criticality Industrial Wireless Networks*, Wireless Networks, https://doi.org/10.1007/978-981-19-8922-3_1

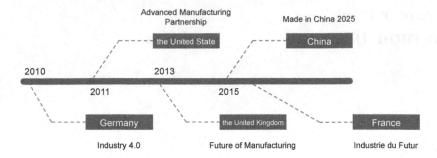

Fig. 1.1 Initiatives of some industrial countries

1.1.1 Network-Based Intelligent Manufacturing Systems

According to a research report by MarketsandMarkets, the global intelligent manufacturing market size in 2021 is USD 88.7 billion, and it is expected to reach USD 228.2 billion by 2027. Some large international companies have turned their attention to intelligent manufacturing and have successfully built related industrial systems. In all of these, the network plays a critical role. Typical systems are as follows:

- *Siemens* provides users with comprehensive intelligent manufacturing solutions from discrete industries to process industries. For example, in the robot collaboration, Siemens' SIMATIC integrator connects robots from different manufacturers through the industrial network PROFIBUS, and then implements the manufacturing process collaboratively. In the automotive industry, Siemens' digital twin that has been applied in the BMW i3 production process adopts the industrial wireless network PROFISAFE to integrate cyberspace and physical space.
- *Boeing* has always been at the forefront of intelligent manufacturing. Based on wide-area networks, it established a global concurrent engineering (GCE) and realized the coordinated development of the upstream and downstream of the industrial chain. This intelligent manufacturing paradigm reduces development costs by 50%.
- *Haier's* intelligent flexible factory allows users on the Internet to participate in the whole process from product design to manufacturing. The factory opened in 2013, and after years of continuous innovation, it now has the ability to meet the needs of mass customization manufacturing. In 2021, its production efficiency has increased by 28%, and the production capacity of a flexible production line is nearly double that of a traditional production line.
- *Schneider Electric* puts forward the concept of green intelligent manufacturing, which integrates energy, automation and the Internet of Things into a holistic system. The system connects data terminals and the cloud management system through the network, so as to realize the high-efficiency and low-power manage-

ment of the whole process of industrial production. The company's ultimate goal is to create a sustainable future industry.
- *Mitsubishi Electric* has built a lithium battery intelligent manufacturing line to support high-precision, high-efficiency and high-flexibility production. The line adopts a time-sensitive network to double the communication performance and increase the production efficiency by 50%.

1.1.2   Industrial Communication Requirements

Since industrial elements must be connected by industrial networks, the quality of service (QoS) of the networks seriously affects industrial production processes. Once the industrial network cannot deliver data packets to their destinations while meeting the communication requirements, it may lead to serious problems such as the decline of control performance, the stagnation of industrial production, and even safety production accidents. Therefore, before designing an industrial network, we must understand the requirements of industrial communications.

In different application scenarios, networks must meet different requirements. The most common requirement is transmission speed, which is present in all networks, not limited to industrial networks. Civilian networks, e.g., WiFi and cellular networks, are constantly evolving to meet users' transmission speed requirements. For this kind of requirement, industrial networks can learn from the advanced techniques of civilian networks to solve the problems in industrial scenarios. In addition to this, industrial networks should meet the following special communication requirements:

- *Real-Time Requirement*: The real-time requirement of communications means that data packets must be delivered to destinations before their deadlines [3]. The real-time requirements in different industrial scenarios are shown in Table 1.1. Safety production has the maximum real-time requirement, and data transmissions have to be completed within 10 ms. Long-delayed data transmissions make it impossible for industrial systems to respond to emergency alarms. This will cause irreparable losses, such as blast furnace explosion, toxic gas leakage, etc. In control-related scenarios, there are some differences in real-time requirements. Closed-loop control has higher real-time requirements, while open-loop control allows more relaxed real-time transmissions. In monitoring-related scenarios, monitoring data do not directly control industrial processes, and log updates even allow a delay of several hours. Therefore, monitoring-related applications have the minimum real-time requirements.
- *Reliability Requirement*: For industrial networks, the reliability requirement means the reception rate of data packets that meet all other communication requirements. The reliability requirements of different industrial scenarios are shown in Table 1.2. Surgical robots have the highest reliability requirement of 99.999999% [4]. If the control commands of the robot are not delivered correctly,

Table 1.1 Real-time requirements

Scenario	Application	Latency
Safety	Emergency shutdown	<10 ms
	Leak detection	
	High-risk process control	
Control	Closed-loop control	10–100 ms
	Automated shutdown	
	Open-loop control	>100 ms
Monitoring	Equipment condition monitoring	
	Environment monitoring	
	System logs	

Table 1.2 Reliability requirements

Scenario	Application	Reliability
Safety	Robotic aided surgery	99.999999%
	Robotic aided diagnosis	99.9999%
Control	Closed-loop control	
	Mobile robots	
	Open-loop control	99.99%
Monitoring	Equipment condition monitoring	99.9%
	Environment monitoring	
	System logs	

it is possible that the patient will be seriously injured. Diagnostic robots get information from bodies without harming the patients, so their reliability can be slightly reduced. In control-related scenarios, the reliability of most applications should reach 99.9999% [5]. However, since the performance of open-loop control is relatively low, their reliability requirements for industrial networks are slightly lower. In monitoring-related scenarios, since monitoring data packets are allowed to transmit for a long time, they can compensate for the low reliability with multiple retransmissions. Therefore, monitoring-related applications have the minimum reliability requirements.

- *Massive Connection Requirement*: Massive connections [6] mean that a large number of network devices are connected to a network at the same time. The number of connected devices around the world shows explosive growth (as shown in Fig. 1.2). In 2003, there were 0.5 billion connected devices around the world, and the number of connected devices per person was 0.08 [7]. Then, in 2020, while the world's population is growing slowly, the number of connected devices is estimated to have increased to 50 billion, and each person has an average of 6.58 connected devices. According to this trend, by 2025, the number of connected devices per person will be 9.27 [8]. The same trend exists in industrial systems. In the early days of industrial networks, only dozens of industrial devices needed to be connected together. Then, with the growth of industrial applications and the improvement of intelligence, the number of industrial devices is increasing exponentially. At present, existing industrial

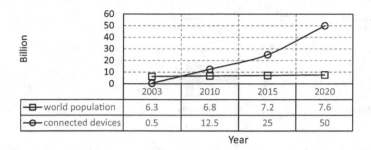

	2003	2010	2015	2020
world population	6.3	6.8	7.2	7.6
connected devices	0.5	12.5	25	50

Year

Fig. 1.2 Massive connection

systems already connect tens of thousands of devices. This trend will continue, and the number of connected devices will be higher in the future.

1.1.3 Classical Industrial Wireless Networks

For the special requirements of industrial communications, industrial wired networks (such as time-sensitive networks, PROFINET, Modbus TCP, etc.) are capable of meeting the real-time and reliability requirements, but cannot connect millions of devices with cables. Therefore, wired networks can only be used in small-scale industrial applications, or only as backbone networks.

Compared with wired networks, wireless networks simplify the network deployment and reduce maintenance costs. Hence, industrial applications should apply wireless networks to connecting devices. However, due to the inherent openness of the wireless environment [9], how to improve the real-time performance and reliability of wireless networks has always been the focus of the industry and scholars. In classical industrial wireless networks, the related techniques and algorithms are as follows:

1.1.3.1 WirelessHART

WirelessHART [10] is a global IEC-approved standard (IEC 62591) that specifies a robust and reliable wireless sensor-actuator network, and has been widely-used in industrial process monitoring and control. WirelessHART is built on top of the IEEE 802.15.4 standard, and defines star and mesh topologies (as shown in Fig. 1.3). It provides enabling techniques for real-time scheduling such as time synchronization, time division multiple access and multiple channels, etc. Although the WirelessHART standard excludes real-time scheduling algorithms, scholars have proposed many related studies. In 2010, the basic idea of multi-processor scheduling was introduced into WirelessHART networks [11]. Thus, in

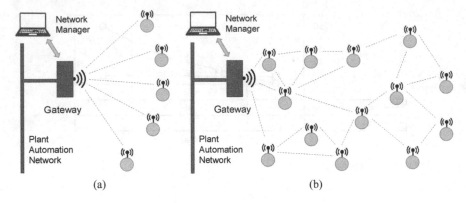

Fig. 1.3 Topologies supported by WirelessHART. (**a**) Star topology. (**b**) Mesh topology

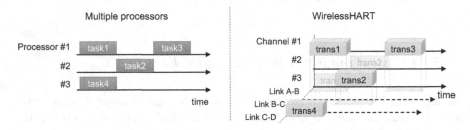

Fig. 1.4 Real-time scheduling model of WirelessHART networks

the scheduling model of WirelessHART networks (as shown in Fig. 1.4), multiple channels correspond to multiple processors, and the time slot used for a transmission corresponds to the running time of a task on processors. However, links and routing paths are excluded in multi-processor systems. Therefore, the main difference between them is that network scheduling must strictly follow the sequence of links in a path, while multi-processor scheduling does not need to consider this. Since the network scheduling model is based on the multi-processor scheduling model, many multi-processor scheduling algorithms, such as fixed priority assignment, dynamic priority scheduling, preemptive scheduling, etc., have been improved to use in WirelessHART networks [12–15].

In addition, to improve the network reliability, WirelessHART adopts the graph routing policy [16–18] (as shown in Chap. 3.2). In a routing graph, there are multiple paths from each node to the destination node, so that data packets can be copied and sent multiple times to improve the reliability. At the same time, WirelessHART supports frequency hopping mechanisms to improve the anti-interference ability of transmissions.

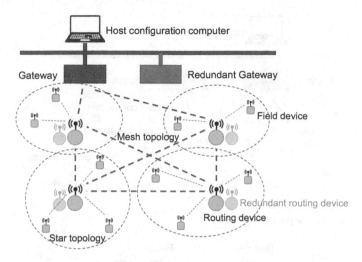

Fig. 1.5 The topology supported by WIA-PA

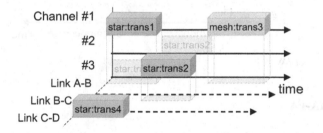

Fig. 1.6 The communication rule of WIA-PA

1.1.3.2 WIA-PA

WIA-PA (IEC 62601) [19] is also based on the IEEE 802.15.4 standard. The main difference from WirelessHART is that WirelessHART defines two topologies (star and mesh), while WIA-PA is a hybrid of star and mesh topologies (as shown in Fig. 1.5). In the star network, field devices upload their data to the routing device. Then, in the mesh network, routing devices transmit data to the gateway. Thus, the communication rule of WIA-PA is specified according to the hybrid topology (as shown in Fig. 1.6). The mesh network cannot transmit until all transmissions in the star network are finished. The hybrid topology can improve the network scalability [20, 21] and enables WIA-PA to support most of the WirelessHART strategies.

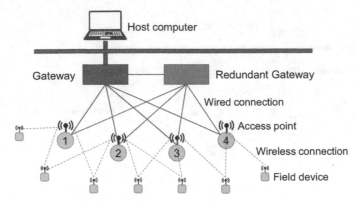

Fig. 1.7 The topology supported by WIA-FA

Fig. 1.8 Segments of WIA-FA

1.1.3.3 WIA-FA

WIA-FA (IEC 62948) [22] and WIA-PA belong to a family of standards. The main difference is that WIA-FA is based on the IEEE 802.11 standard. Hence, WIA-FA only defines a redundant star topology (as shown in Fig. 1.7) and does not need any routing mechanism. To improve the real-time performance and reliability, WIA-FA adopts multiple access points and redundant devices. Based on these features, the communication of WIA-FA is segmented (as shown in Fig. 1.8). Access points are divided into several groups, and the transmission time is correspondingly divided into the same number of segments. Each group is assigned one segment. The access points of a group are allowed to occupy only the assigned segment. In a segment, first, access points broadcast beacons to field devices, and then according to the configuration in beacons, field devices start to work. Since the topology of WIA-FA is a simplification of WIA-PA and WirelessHART, some of the algorithms of WIA-PA and WirelessHART can be used in WIA-FA. In addition, based on multiple access points, a real-time scheduling algorithm for highly reliable retransmissions has been proposed to meet industrial communication requirements [23].

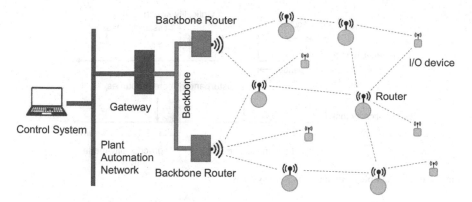

Fig. 1.9 The topology supported by ISA100.11a

1.1.3.4 ISA100.11a

ISA100.11a (IEC 62734) [24] is also based on the IEEE 802.15.4 standard. Its topology is similar to that of WirelessHART (as shown in Fig. 1.9), and most of the techniques used in WirelessHART to improve the real-time performance and reliability also exist in ISA100.11a, such as time division multiple access, multiple channels and graph routing. Therefore, the scheduling model and relevant strategies of WirelessHART can also be applied to ISA100.11a. In addition, the design goal of ISA100.11 is different from those of the other networks. ISA100.11a was designed for ease of use. Flexible configuration interfaces and IPv6 are included in ISA100.11a. These techniques enable ISA100.11a to easily connect to other networks.

1.2 Mixed Criticality

Mixed criticality [25] is a new system characteristic, which is different from the real-time performance and reliability of systems. However, the methods proposed based on mixed criticality can guarantee the real-time performance and reliability of the systems with limited resources. In addition, since the mixed-criticality methods are aimed at limited resources, they can also be a solution to support massive connections.

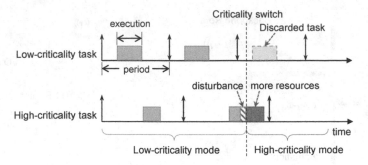

Fig. 1.10 An example of mixed-criticality systems

1.2.1 Mixed Criticality Overview

Mixed-criticality systems define extensions to real-time embedded systems with limited resources. The traditional real-time performance measures the urgency of tasks, not their importance. Then, mixed-criticality systems are proposed to highlight the importance of tasks. In such systems, important tasks are called high-criticality tasks, and unimportant tasks are called low-criticality tasks. Mixed-criticality systems can execute in multiple criticality modes, and criticality level is also one of the attributes of tasks. To make it clearer, in the following, a system model with two criticality levels (high and low) is introduced. For multiple criticality levels, the system model can be easily extended.

As shown in Fig. 1.10, when the system with two criticality levels starts, it must be in the low-criticality mode, and all tasks are allowed to execute. During system execution, some internal or external dynamic disturbances may cause high-criticality tasks to fail to meet performance requirements. Then, the system switches to the high-criticality mode. In the high-criticality system mode, high-criticality tasks are allowed to use more resources to guarantee their real-time performance and reliability, while low-criticality tasks have to be discarded if there are not sufficient resources. In this way, the performance of high-criticality tasks can be guaranteed as much as possible, and then the system performance can be maximized under limited resources.

For different systems, mixed-criticality models are different [26]. In this book, we only focus on industrial wireless networks.

1.2.2 Mixed Criticality in Industrial Wireless Networks

In industrial wireless networks, criticality is an inherent property of data packets [27, 28]. For example, Fig. 1.11 shows an industrial wireless network for cement manufacturing. The rotary kiln is the most important equipment. If the rotary kiln

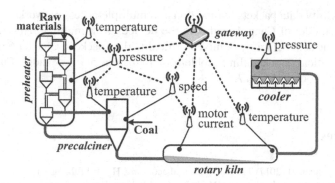

Fig. 1.11   An industrial wireless network in a cement factory

has some exceptions and its temperature data are lost or miss deadline, workers cannot take measures in time. This will lead to production inefficiency. By contrast, even if the temperature data of pre-heaters cannot be delivered to the destination within deadlines, the temperature of materials can be sensed in the pre-calciner. Hence, the temperature data of the rotary kiln are more critical than those of pre-heaters. In the wireless network, when the important equipment has an exception, its sensing data have to be quickly and reliably delivered to the control room. This process needs more network resources. However, the resources of wireless networks are very limited. When the resources cannot guarantee the requirements of all data packets, the packets that belong to lower-criticality levels have to be discarded.

Based on the above basic rules of mixed-criticality wireless networks, some novel algorithms are needed to analyze and schedule the mixed-criticality packets so that as many packets as possible can be satisfied under performance constraints. The key feature of mixed-criticality systems is that the resources assigned to high-criticality tasks can be dynamically increased during system execution. Therefore, the main difference among the various mixed-criticality systems is what the resources are and how to use the resources. For example, in mixed-criticality multi-processor systems, the resource is the execution time of a task on processors, while, in mixed-criticality wireless networks, there are various types of resources, including time slots and channels at the data link layer, and routing path at the network layer. In this book, we will introduce some algorithms to solve the analyzing and scheduling problem of mixed-criticality data packets from the perspective of network resources.

1.3   Book Organization

In this book, we will see how mixed criticality helps to improve the performance of industrial wireless networks. The rest of this book is organized as follows: In Chaps. 2 and 3, we analyze the schedulability of mixed-criticality data packets under two classical scheduling algorithms. Chapter 4 presents a new algorithm to schedule mixed-criticality data packets in time-division multiple-access networks. Chapter 5 extends the model of Chap. 4, and uses the devices with multiple radio interfaces to improve the schedulability of mixed-criticality data packets. Chapter 6 addresses the mixed-criticality scheduling problem of industrial 5G new radio. Finally, we conclude this book in Chap. 7.

References

1. Neiburg F, Guyer JI (2017) The real in the real economy. HAU: J Ethnog Theory 7(3):261–279
2. Montazerolghaem A, Yaghmaee MH (2020) Load-balanced and QoS-aware software-defined Internet of Things. Internet Things J 7(4):3323–3337
3. Jin X, Xia CQ, Xu C (2022) Real-time network scheduling for industrial all-element interconnection: progress and challenges. Commun CCF 18(8):24–28
4. Tan J, Sha XB, Dai B, Lu T (2021) Analysis of industrial Internet of Things and digital twins. ZTE Commun 19(2):53–60
5. Boban M, Giordani M, Zorzi M (2021) Predictive quality of service: the next frontier for fully autonomous systems. IEEE Netw 35(6):104–110
6. Zhan W, Xu C, Sun X, Zou J (2021) Toward optimal connection management for massive machine-type communications in 5G system. Internet Things J 8(17):13237–13250
7. Calsoft (2018) Internet of Things (IoT) 2018–Market Statistics, Use Cases and Trends. https://futureiot.tech/wp-content/uploads/2018/11/Calsoft-IoT-2018-Market-Statistics-Use-Cases-and-Trends.pdf
8. Safaei B, Monazzah AMH, Bafroei MB (2017) Reliability side-Effects in Internet of things application layer protocols. In: International conference on system reliability and safety (ICSRS). IEEE, pp. 207–212
9. Zhou F, Feng L, Kadoch M, Yu P, Li W, Wang Z (2021) Multiagent RL aided task offloading and resource management in Wi-Fi 6 and 5G Coexisting industrial wireless environment. Trans Ind Informat 18(5):2923–2933
10. Song J, Han S, Mok A, Chen D, Lucas M, Nixon M, Pratt W (2008) WirelessHART: applying wireless technology in real-time industrial process control. In: Real-time and embedded technology and applications symposium (RTAS). IEEE, pp 377–386
11. Saifullah A, Xu Y, Lu C, Chen Y (2010) Real-time scheduling for WirelessHART networks. In: Real-time systems symposium (RTSS). IEEE, pp 150–159
12. Nobre M, Silva I, Guedes LA (2015) Routing and scheduling algorithms for WirelessHART networks: a survey. Sensors 15(5):9703–9740
13. Modekurthy VP, Saifullah A, Madria S (2019) DistributedHART: a distributed real-time scheduling system for WirelessHART networks. In: Real-time and embedded technology and applications symposium (RTAS). IEEE, pp 216–227
14. Chen G, Cao X, Liu L, Sun C, Cheng Y (2018) Joint scheduling and channel allocation for end-to-end delay minimization in industrial WirelessHART networks. Internet Things J 6(2):2829–2842
15. Saifullah A, Xu Y, Lu C, Chen Y (2011) Priority assignment for real-time flows in WirelessHART networks. In: Euromicro conference on real-time systems (ECRTS). IEEE, pp 35–44
16. Dang K, Shen JZ, Dong LD, Xia YX (2013) A graph route-based superframe scheduling scheme in WirelessHART mesh networks for high robustness. Wirel Pers Commun 71(4):2431–2444
17. Modekurthy VP, Saifullah A, Madria S (2018) Distributed graph routing for WirelessHART networks. In: The international conference on distributed computing and networking, pp 1–10

18. Shi J, Sha M, Yang Z (2018) DiGS: distributed graph routing and scheduling for industrial wirelss sensor-actuator networks. In: The international conference on distributed computing systems (ICDCS). IEEE, pp 354–364
19. Liang W, Zhang XL, Xiao Y, Wang F, Zeng P, Yu HB (2011) Survey and experiments of WIA-PA specification of industrial wireless network. Wirel Commun Mob Comput 11(8):1197–1212
20. Wang H, Chen PF, Wang P (2018) Deterministic scheduling algorithms for WIA-PA industrial wireless sensor networks. Acta Electon Sin 46(1):68–74
21. Jin X, Xu HT, Xia CQ, Wang JT, Zeng P (2018) Convergecast scheduling and cost optimization for industrial wireless sensor networks with multiple radio interfaces. Wirel Netw 24:3205–3219
22. Liang W, Zheng M, Zhang J, Shi H, Yu H, Yang Y, Zhao X (2019) WIA-FA and its applications to digital factory: a wireless network solution for factory automation. Proc IEEE 107(6):1053–1073
23. Shi H, Zheng M, Liang W, Zhang J (2021) AODR: an automatic on-demand retransmission scheme for WIA-FA networks. Trans Veh Technol 70(6):6094–6107
24. Nixon M, Rock TR (2012) A comparison of WirelessHART and ISA100.11a. Whitepaper, Emerson Process Management, pp 1–36
25. Vestal S (2007) Preemptive scheduling of multi-criticality systems with varying degrees of execution time assurance. In: Real-time systems symposium (RTSS). IEEE, pp 239–243
26. Burns A, Davis RI (2022) Mixed criticality systems - a review. https://www-users.cs.york.ac.uk/burns/review.pdf
27. Harbin JR, Griffin DJ, Burns A, Bate IJ, Davis RI, Indrusiak LS (2018) Supporting critical modes in AirTight. In: Workshop on mixed criticality, pp 7–12
28. Burns A, Harbin JR, Indrusisk LS, Bate IJ, Davis RI, Griffin DJ (2018) AirTight: A resilient wireless communication protocol for mixed criticality systems. In: The international conference on embedded and real-time computing systems and applications (RTCSA). IEEE, pp 65–75

Chapter 2
Schedulability Analysis of Mixed-Criticality Data Under Fixed-Priority Scheduling

Abstract WirelessHART, as a robust and reliable wireless protocol, has been widely-used in industrial systems. Its real-time performance has been extensively studied, but limited to the single-criticality case. Many advanced applications have mixed-criticality communications, where different data flows come with different criticality levels. Hence, in this chapter, we study the real-time mixed-criticality communication based on WirelessHART networks, and present an end-to-end delay analysis method under fixed priority scheduling.

2.1 Background

WirelessHART is based on a centralized network management and multi-channel Time Division Multiple Access (TDMA). These special features have attracted researchers' attentions, and they have done some studies to improve the real-time performance of WirelessHART networks, e.g. [1–4]. However, all these studies focus on the single-criticality case. Advanced real applications come with mixed-criticality data communications, such as the case of the cement factory in Sect. 1.2.2.

The key difference between mixed- and single-criticality systems is that the criticality of data in mixed-criticality systems must be considered together with real-time performance [5]. This leads to the problem that directly using traditional priority-based scheduling algorithms of single-criticality systems to mixed-criticality systems is infeasible due to independence between criticality and traditional priorities [6–12]. Therefore, the traditional real-time theory needs a revision to support mixed-criticality networks.

There are a few related studies on mixed-criticality networks. The work in [13–15] designs Network-on-Chips for mixed-criticality multiprocessor systems. The work in [16] proposes mixed-criticality protocols for the Controller Area Network (CAN), and then a response-time analysis method and an optimal priority assignment scheme are provided. The work in [17] designs a virtual CAN controller

© The Author(s) 2023
X. Jin et al., *Mixed-criticality Industrial Wireless Networks*, Wireless Networks,
https://doi.org/10.1007/978-981-19-8922-3_2

to provide differentiated services for different criticality levels. The work in [18–20] focuses on the wired network—TTEthernet. They propose some scheduling algorithms to guarantee the performance of messages under real-time constraints. The work in [21] is about wireless networks. It introduces a mixed-criticality scheduling method to JPEG2000 video systems based on the IEEE 802.11 standard. However, WirelessHART networks are based on the IEEE 802.15.4 standard and are quite different from the wireless video system. Therefore, existing system models and solutions cannot be used in the WirelessHART model.

Since the end-to-end delay analysis is the foundation of the real-time theory, in this chapter, we present an end-to-end delay analysis method for fixed priority scheduling in mixed-criticality real-time WirelessHART networks. The analysis can be used to test whether the data flows can meet their special requirements when designing a WirelessHART network.

In the following, first, we introduce the concept of mixed criticality into real-time wireless sensor-actuator networks and propose a formulated system model; second, we propose an end-to-end delay analysis method, which is a fast feasible method to test the reliability of mixed-criticality systems; third, evaluation results show that the proposed method is very effective and only incurs little pessimism comparing with simulation results and a real testbed.

2.2 System Model

2.2.1 Mixed-Criticality Wireless Network Model

We consider a WirelessHART network characterized by $G = < V, E, m >$:

- A WirelessHART network consists of sensor/actuator nodes and a gateway with a centralized network manager. We use n nodes $V = \{v_1, v_2, \ldots, v_n\}$ to denote these devices, and the gateway is v_1. Each node is equipped with a transmitter, so it cannot send and receive in the same time slot.
- $E : V \times V$ is the set of links. Each element e_{ij} in E represents existing reliable communication between v_i and v_j. Transmitting a packet through one link is called *transmission*.
- We use m to denote the number of available channels. WirelessHART networks support 16 non-overlapping channels. However, since these channels may suffer from persistent external interference, not all of them can always be accessed. Hence, $0 < m \leq 16$. Each channel supports only one transmission in one time slot.

The data flow set is denoted by $\mathbb{F} = \{F_1, F_2, \ldots\}$. Each flow F_i is characterized by $F_i = < \chi_i, c_i, t_i(x), p_i, \phi_i >$. p_i denotes the distinct fixed priority. ϕ_i ($\phi_i \subseteq E$) is an ordered sequence of links and denotes the routing path of the flow F_i. The centralized manager of the WirelessHART network collects sensing data and distributes actuator data, so the gateway is the source or destination for each flow. In the TDMA policy, each time slot allows a one-hop data transmission and its acknowledgement to be transmitted. We use c_i to denote the number of time slots required to deliver a packet from the source to the destination, i.e., c_i is equal to the number of hops of the flow F_i.

χ_i denotes the criticality level of the flow F_i. For ease of presentation, we only focus on a dual-criticality system, in which there are only two criticality levels L (low) and H (high). However, it can be easily extended to systems with an arbitrary number of criticality levels. Correspondingly, the network also has dual-criticality modes $\{H, L\}$. If the criticality level of the flow F_i is not less than the current network mode x, it can be delivered; otherwise, the flow is discarded. The network starts in the low-criticality mode ($x = L$), in which all flows are served. When an error or exception occurs in a node, the node will trigger the changing of the network mode from low criticality to high criticality ($x = H$). Then only the flows with high-criticality level can be delivered, and the low-criticality flows are discarded. Note that the mode change will introduce additional time to the delay of the high-criticality flow and the message of mode change should be broadcast to the entire network as soon as possible. There are some methods used to solve this problem. For example, one channel of each node can be reserved to serve the message. Therefore, we only model the duration of the mode change as C, which is used to calculate the delay of the packet delivered during the mode change.

When errors and exceptions occur, workers will handle problems and change the mode from high criticality to low criticality. We do not consider this process due to the unpredictability of workers' behavior, i.e., we do not study the mode change from high criticality to low criticality. The assumption is also widely adopted in existing works (such as [22–24] etc.).

In mixed-criticality uniprocessor/multiprocessor systems, the execution time of a task is a function of the system mode. In wireless networks, the number of time slots required to deliver a packet is equal to the number of hops and fixed. However, the period t_i is dependent on the network criticality mode. Since the important flow is more frequently delivered when the network mode is changed to high criticality, hence, $t_i(H) < t_i(L)$.

According to the period $t_i(x)$, the flow F_i periodically releases a packet, which is assigned the parameters specified in F_i. Our system adopts the implicit-deadline, i.e., the packet's relative deadline is equal to the flow's period corresponding with the network mode of generating the packet. For example, if the packet is released in the network mode L, then its relative deadline is $t_i(L)$. Therefore, in a stable network, at most one packet of each flow is active at any time. However, when the network mode is changed from L to H, there may exist two active packets belonging to one flow because of the change of the flow's period. In this case, the packet released in the network mode H has higher priority than another.

A packet is released at time slot s_1, and arrives at its destination at time slot s_2, then its end-to-end delay is $(s_2 - s_1 + 1)$. The end-to-end delay of a flow is the maximum delay among all its packets. If a scheduling algorithm can schedule all flows such that all packets' end-to-end delays are less than or equal to their deadlines, the flow set is called *schedulable* under the scheduling algorithm.

Note that not all of the above assumptions are supported by the original WirelessHART protocol. However, they can be implemented in the application layer [1–3]. In Sect. 2.4, our real testbed is introduced.

2.2.2 Fixed Priority Scheduling

We focus on the end-to-end delay analysis for fixed priority scheduling, which is the most commonly used real-time scheduling in real systems. In fixed priority scheduling, transmissions are scheduled within a hyper-period T, which is equal to the least common multiple of periods of all flows, since after that all schedules are cyclically repeated. The period supported by the WirelessHART protocol is 2^i, where i is an integer greater than or equal to 0. Therefore, the hyper-period T is equal to the maximum period among all flows. At each time slot of T, if there exist unused channels, the transmission with the highest priority is scheduled. However, if the released transmission shares nodes with the transmissions that have been scheduled at this time slot, it cannot be scheduled since one node cannot serve more than one transmission at one time slot (as shown in Fig. 2.1). Therefore, there are two factors influencing the transmission scheduling: *channel contention* (there are no unused channels assigned to the transmission) and *transmission conflicts* (a transmission cannot be scheduled, if it shares a node with a transmission that has been scheduled in this time slot). In other words, the two factors introduce extra delays. In the following, we analyze delays introduced by two factors separately.

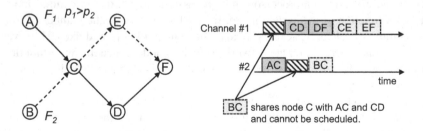

Fig. 2.1 Fixed priority scheduling

Table 2.1 Key notations

Symbol	Definition
G	A WirelessHART network
V	Set of all devices in the network G
E	Set of links in the network G
m	Number of available channels
F_i	A data flow, $F_i \in \mathbb{F}$
χ_i	Criticality level of the flow F_i
c_i	Number of hops of the flow F_i
x	Criticality mode of the current network
$t_i(x)$	Period of the flow F_i at the current mode
p_i	Priority of the flow F_i
ϕ_i	Routing path of the flow F_i
C	Duration of the mode change
s_i	The ith time slot
T	Hyper-period
R_k^{ch}	Pseudo worst case delay of the flow F_k for single networks
$hp(F_k)$	Set of flows whose priorities are higher than F_k
$R_k(x)$	Worst case delay of the packet that belongs to the flow F_k at the network mode x
r	Number of hops that the packet has passed before the mode change

Problem Statement Given the mixed-criticality WirelessHART network G, the flow set $\mathbb{F}$ and the fixed priority scheduling algorithm, our objective is to analyze the end-to-end delay for each flow, such that the schedulability of the flow set can be determined.

Table 2.1 summarizes the key notations used in this chapter.

2.3 End-to-End Delay Analysis

Our analysis is based on the *EDA* (End-to-end Delay Analysis) method [1], which is the state-of-the-art end-to-end delay analysis for fixed priority scheduling in single-criticality real-time WirelessHART networks. To make this chapter self-contained, we first introduce EDA. Then we present our end-to-end delay analysis for mixed-criticality WirelessHART Networks.

2.3.1 Analysis for Single-Criticality Networks

The EDA analysis contains two steps. The first step calculates the delay due to channel contention, which is called *pseudo upper bound* of the worst case end-to-end delay and denoted by R_k^{ch}. Then the second step incorporates the delay due to transmission conflicts into the result of the first step.

2.3.1.1 Pseudo Delay

The flow F_k experiences the worst case delay when the level-k busy period occurs. The *level-k busy period* is the maximum continuous time interval during which all channels are occupied by flows of priority higher than the priority of F_k, until F_k finishes its active packet transmitting. The notation $hp(F_k)$ is used to denote the set of flows whose priorities are higher than F_k. If the flow F_i ($F_i \in hp(F_k)$) has a packet with release time earlier than the level-k busy period and deadline in the level-k busy period, it is said to have *carry-in* workload in the busy period. Then two types of workload are presented as follows:

- $W_k^{NC}(F_i, \alpha)$ denotes the workload upper bound in the level-k busy period of α slots, if F_i has no carry-in workload:

$$W_k^{NC}(F_i, \alpha) = \left\lfloor \frac{\alpha}{t_i} \right\rfloor \cdot c_i + \min(\alpha \bmod t_i, c_i)$$

where t_i denotes the period of F_i in single-criticality networks.

- $W_k^{CI}(F_i, \alpha)$ denotes the workload upper bound in the level-k busy period of α slots, if F_i has a carry-in workload:

$$W_k^{CI}(F_i, \alpha) = \left\lfloor \frac{\max(\alpha - c_i, 0)}{t_i} \right\rfloor \cdot c_i + c_i + \mu_i$$

where $\mu_i = \min(\max(\max(\alpha - c_i, 0) - (t_i - R_i), 0), c_i - 1)$ and R_i denotes the worst case end-to-end delay of F_i in single-criticality networks.

Similarly, there are two types of interference between F_i and F_k during α slots:

$$I_k^{NC}(F_i, \alpha) = \min(W_k^{NC}(F_i, \alpha), \alpha - c_k + 1) \tag{2.1}$$

$$I_k^{CI}(F_i, \alpha) = \min(W_k^{CI}(F_i, \alpha), \alpha - c_k + 1) \tag{2.2}$$

At most $m - 1$ higher priority flows have carry-in workload in the network with m channels. Therefore, F_k's total delay due to channel contention is

$$\Omega_k(\alpha) = \sum_{F_i \in hp(F_k)} I_k^{NC}(F_i, \alpha) + U_k(\alpha)$$

where $U_k(\alpha)$ is the sum of the min($|hp(F_k)|$, $m-1$) largest values of the differences $I_k^{CI}(F_i, \alpha) - I_k^{NC}(F_i, \alpha)$ among all $F_i \in hp(F_k)$.

The WirelessHART network contains m channels, so Eq. (2.3) shows the delay due to channel contention. And the pseudo upper bound R_k^{ch} is the minimal value of α that solves Eq. (2.3). α can be found using the iterative fixed-point algorithm [25], which is widely used in the delay analysis of real time systems. The iterative calculation of α starts at $\alpha = c_k$. During the iterations, if α is larger than the deadline of the flow F_k, the algorithm terminates and the flow set is unschedulable; if the value of α is fixed and less than the deadline, the fixed-point is R_k^{ch}.

$$\alpha = \left\lfloor \frac{\Omega_k(\alpha)}{m} \right\rfloor + c_k \qquad (2.3)$$

2.3.1.2 Worst Case Delay

This step incorporates the delay due to transmission conflicts into R_k^{ch} to calculate the actual end-to-end delay R_k. First, we introduce some definitions.

- $Q(k, i)$: the total number of F_i's transmissions that share nodes with F_k's transmissions.
- $\delta_j(k, i)$: the number of nodes along the jth maximal common path between F_k and F_i. $\delta_j'(k, i)$ is the length of the maximal common path with a length of at least 4. The delay caused by a maximal common path is at most 3, so the extra length is specially marked using $\delta_j'(k, i)$.
- $\Delta(k, i)$: the upper bound of end-to-end delay due to transmission conflicts that F_i contributes to F_k,

$$\Delta(k, i) = Q(k, i) - \sum_{j=1}^{\sigma} (\delta_j'(k, i) - 3)$$

where σ is the number of maximal common paths between F_k and F_i.

Thus the upper bound of the actual delay R_k is the minimal solution of Eq. (2.4) by running the iterative fixed point algorithm starting at $\beta = R_k^{ch}$.

$$\beta = R_k^{ch} + \sum_{F_i \in hp(F_k)} \left\lceil \frac{\beta}{t_i} \right\rceil \cdot \Delta(k, i) \qquad (2.4)$$

2.3.2 Analysis for Mixed-Criticality Networks

Mixed-criticality networks dynamically change the network mode, which results in three types of packets transmitted in the network:

- The release time and deadline of a packet are all in the network mode L. The end-to-end delay of this packet is denoted by $R_k(L)$.
- The release time and deadline of a packet are all in the network mode H. The notation $R_k(H)$ is used to present the upper bound of its delay.
- When the network mode is changed, the packet, which is released by high-criticality flow in the network mode L, cannot be dropped. In this situation, the packet's release time is in the network mode L, but its deadline is in the H mode. Flows formed by these packets are delivered only once. The notation $\mathbb{F}'$ presents the set of these flows and $R_k(L2H)$ denotes the upper bound of the delay.

$R_k(L)$ is unaffected by the mode change and is equal to the delay R_k calculated in single-criticality networks. Therefore, we only analyze $R_k(H)$ and $R_k(L2H)$.

2.3.2.1 Analyzing $R_k(H)$

In the network mode H, packets belonging to the high-criticality flow are delivered, no matter when they are released. Therefore, $R_k(H)$ is interfered by the following two flow sets

$$hpL(F_k) = \{F_i | F_i \in \mathbb{F}', p_i < p_k, \chi_i = H\},$$

$$hpH(F_k) = \{F_i | F_i \in \mathbb{F}, p_i > p_k, \chi_i = H\}.$$

From these, we can derive that the delay due to channel contention is

$$\Omega_k(\alpha) = \sum_{F_i \in hpH(F_k) \cap hpL(F_k)} I_k^{NC}(F_i, \alpha) + U_k(\alpha)$$

where $U_k(\alpha)$ is also for the interferences of $hpH(F_k)$ and $hpL(F_k)$. Note that the interferences of flows in $hpH(F_k)$ are the same with Eqs. (2.1) and (2.2), since the flows release packets periodically. However, the flow F_i ($F_i \in hpL(F_k)$) is delivered only once. In the worst case, its workload in the level-k busy period is $\min\{\alpha, c_i\}$. Therefore,

$$\forall F_i \in hpL(F_k) : I_k^{CI}(F_i, \alpha) = I_k^{NC}(F_i, \alpha) = \min(\min(\alpha, c_i), \alpha - c_k + 1)$$

Then the pseudo upper bound $R_k^{ch}(H)$ can be derived based on Eq. (2.3).

For the delay due to transmission conflicts, similarly, besides the periodic flows in $hpH(F_k)$, the flows in $hpL(F_k)$ delivered only once will introduce $\Delta(k, i)$ to

$R_k(H)$. Therefore,

$$\beta = R_k^{ch}(H) + \sum_{F_i \in hpH(F_k)} \left\lceil \frac{\beta}{t_i(H)} \right\rceil \cdot \Delta(k, i) + \sum_{F_i \in hpL(F_k)} \Delta(k, i) \qquad (2.5)$$

According to Eq. (2.5), the iterative algorithm can be used to find the fixed β, i.e., $R_k(H)$.

2.3.2.2 Analyzing $R_k(L2H)$

The flow F_k ($F_k \in \mathbb{F}'$) is divided into two flows. The first flow F_{kL} is delivered in the L mode, and the second flow F_{kH} is in the H mode. We use $R_k^r(L)$ and $R_k^r(H)$ to denote the delays of F_{kL} and F_{kH}, respectively, where r means that the packet has passed through r hops before the mode change, and $r \in [0, c_k - 1]$. Correspondingly, $c_{kL} = r$ and $c_{kH} = c_k - r$. And priorities of F_{kL} and F_{kH} are assigned as p_k.

The calculation of $R_k^r(L)$ is the same as that of $R_k(L)$, since they are all in the stable network. However, $R_k^r(H)$ is different from $R_k(H)$. According to our system model, packets released by F_k in the network mode H have higher priority than the packets of F_{kH}. Therefore, the delay contributed by these higher priority packets must be added to $R_k^r(H)$, i.e., F_{kH} is interfered by $hpL(F_k)$, $hpH(F_k)$ and $\{F_k\}$ in the network mode H. From these, we can derive

$$\Omega_{kH}(\alpha) = \sum_{F_i \in hpH(F_k) \cap hpL(F_k) \cap \{F_k\}} I_{kH}^{NC}(F_i, \alpha) + U_{kH}(\alpha)$$

where U_{kH} is for $hpL(F_k)$, $hpH(F_k)$ and $\{F_k\}$. For the flow F_{kH}, F_k releases higher priority packets periodically. The interference introduced by F_k is the same as that by $hpH(F_k)$. Therefore, Eqs. (2.1) and (2.2) also can be used to calculate it.

For the delay due to transmission conflicts, the packets released by $\{F_k\}$ must be considered. Then the actual end-to-end delay is shown as follows:

$$\beta = R_{kH}^{ch}(H) + \sum_{F_i \in hpH(F_k) \cap \{F_k\}} \left\lceil \frac{\beta}{t_i(H)} \right\rceil \cdot \Delta(kH, i) + \sum_{F_i \in hpL(F_k)} \Delta(kH, i)$$

And $R_k^r(H)$ is also solved by the iterative algorithm.

The range of r is from 0 to $c_k - 1$. If $r = 0$, it means that the packet has been released but not been delivered before the network mode is changed to H. Thus, $c_{kL} = 0$. This will cause the failure of the iterative algorithm. Therefore, if $\exists F_i$ and $p_i > p_k$, α starts with 1; otherwise, there is no interference for F_k and $R_k^r(L) = 0$. If $r = c_k$, it means that the packet has been delivered to its destination in the L network mode. Hence, the delay of the packet is $R_k(L)$.

The delay of F_k is the sum of $R_k^r(L)$ and $R_k^r(H)$. However, different values of r lead to different $R_k^r(L)$ and $R_k^r(H)$. Therefore, the upper bound of F_k's delay is

$$R_k(L2H) = \max_{r \in [0, c_k - 1]} \{R_k^r(L) + R_k^r(H)\} + C$$

where C is the additional time introduced by the mode change (shown in Sect. 2.2).

To sum up, if the flow F_k satisfies $R_k(L) \leq t_k(L)$, $R_k(H) \leq t_k(H)$ and $R_k(L2H) \leq t_k(L)$, then it is schedulable. In a flow set, if all the flows are schedulable, the flow set is schedulable. The calculation of our analysis is in pseudo polynomial time because our analysis is based on the iterative fixed-point algorithm.

2.4 Performance Evaluations

In this section, we will compare our analysis method with simulations and a real testbed.

2.4.1 Simulations

In order to illustrate the applicability of our method, for each parameter configuration, 100 test cases are generated randomly. For each test case, the gateway is placed at the center and other nodes are placed randomly in the playground area A. According to the suggestion in [26], the number of nodes n and the playground area A should satisfy

$$\frac{n}{A} = \frac{2\pi}{d^2 \sqrt{27}}$$

where the transmitting range d is set as $40\,\mathrm{m}$. Then, each node connects to the nearest node, which must be in its transmitting range and has been connected to the gateway. If some nodes cannot connect to the gateway, their locations are generated randomly again.

The flow set $\mathbb{F}$ contains $0.8 \cdot n$ flows. Other parameters are set as follows. We use the utilization u ($u = \sum_{\forall F_i \in \mathbb{F}} c_i/t_i$) to control the workload of the entire network, and *UUniFast algorithm* [27] is used to generate each flow's utilization u_i ($u_i = c_i/t_i$). The result generated by UUniFast algorithm follows a uniform distribution and is neither pessimistic, nor optimistic for the analysis. For the flow F_i, its criticality level is assigned randomly. If $\chi_i = H$, then $t_i(L) = 2^{\lceil \log_2^{ti} \rceil}$ and $t_i(H) = 2^{\lfloor \log_2^{ti} \rfloor}$; otherwise, $t_i(L) = 2^{\lceil \log_2^{ti} \rceil}$ and $t_i(H) = +\infty$. The mode change duration C is set as the maximum number of hops between any two nodes of the network. If one channel and one transmitter of each node are reserved to serve the mode change, the change command can be broadcast to all the nodes in the duration C. The fixed priority assignment follows the two classical algorithms [28]: (1) Deadline Monotonic (DM), in which the flow with the shorter deadline is assigned the higher priority; (2) Proportional Deadline monotonic (PD), in which the flow with the shorter subdeadline is assigned the higher priority. *Subdeadline* is defined for its deadline divided by the total number of its transmissions.

The mode change can occur at any time slot. Hence, the simulation should list all cases. However, for the complex state space, the execution time of simulations is unacceptable. Therefore, if the execution time exceeds 30 minutes, the simulation is suspended and the maximum delay is chosen as the worst case end-to-end delay. We use *pessimism ratio* (the proportion of our analyzed delay to the maximum delay observed in simulations) and *acceptance ratio* (the percentage of flow sets that are schedulable) as the performance metrics.

Figure 2.2 plots the pessimism ratios with different numbers of nodes. We set that $m = 12$ and $u = 1$. In order to make test cases simulated in an acceptable time, the number of nodes is only up to 110. From the figures, we can see that the 75th percentile of the pessimism ratios is less than 2.1 and 2.2 for DM and PD, respectively. In [1], the result of the state-of-the-art analysis EDA for single-criticality networks is 1.5 and 1.6, respectively. Compared with them, our analysis only introduces a small degree of pessimism, even though the mode change increases the complexity of the end-to-end delay analysis. Therefore, our analysis is highly effective.

In order to evaluate the performance of our analysis method for the larger scale networks in an acceptable time, we set $m = 6$ and $u = 1$. Figure 2.3 shows the boxplots of the pessimism ratios under varying network sizes. From the evaluation results, we know that our analysis method is stable under different network sizes. Comparing with Fig. 2.2, Fig. 2.3 is more pessimistic. The less number of channels introduces more contentions, and the delay analysis is to consider the worst case scenario. Thus, all of additional contentions are considered in the delay analysis, but not all of them appear in simulations. Therefore, the analysis with fewer channels is more pessimistic.

We compare the acceptance ratios of our analysis and simulations, and the utilization u is increased to 3.2. Figure 2.4 shows the comparison, in which AMC is our Analysis for Mixed-Criticality networks and SIM is the result of simulations. We observe that our results are close to those of simulations. Therefore, our analysis method can be used to verify whether flows can meet their deadlines or not before implementing the real system.

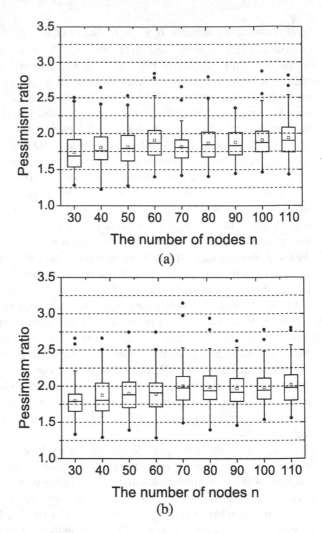

Fig. 2.2 Pessimism ratio under varying network sizes with $m = 12$. (a) The priority assignment policy DM. (b) The priority assignment policy PD

Comparing Fig. 2.4a and b, we observe that the acceptance ratio of the policy PD is less than that of the policy DM. It is because that, compared with the policy DM, the policy PD introduces more interferences to the flows with short paths, which leads to a longer delay. Similarly, all the interferences are considered in the delay analysis, but not all of them appear in simulations. Therefore, in Figs. 2.2 and 2.3, the result of PD is more pessimistic than that of DM.

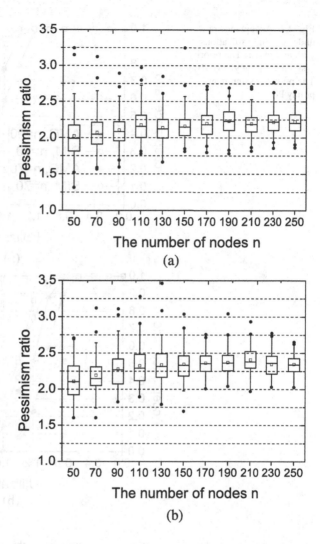

Fig. 2.3 Pessimism ratio under varying network sizes with $m = 6$. (**a**) The priority assignment policy DM. (**b**) The priority assignment policy PD

2.4.2   Real Testbed

We implement a real testbed that contains three types of physical devices: the gateway device, routing devices and field devices. The gateway device manages the network and adopts a low power SoC (System of Chip) AT91RM9200 and a CC2420 transceiver chip. The routing device is implemented on an MSP430 and a CC2420. The field device is equipped with a temperature and humidity sensor SHT15 besides an MSP430 and a CC2420. Our testbed supports the IEEE 802.15.4 protocol, which is the physical and MAC (Medium Access Control) layers of Wire-lessHART networks, and an improved WirelessHART network according to our requirements. The improved WirelessHART implements the specific requirements

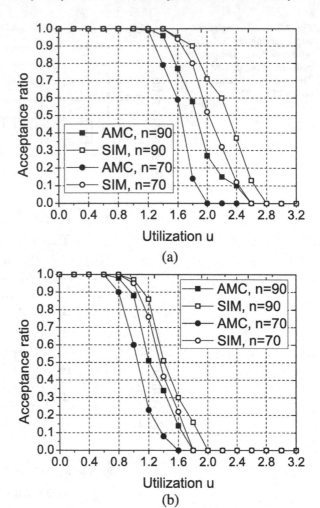

Fig. 2.4 Acceptance ratio under varying utilizations with $m = 6$. (**a**) The priority assignment policy DM. (**b**) The priority assignment policy PD

in the application layer, and it is compatible with the original WirelessHART. Channel 23 is used to broadcast mode change messages and configuration messages. Six schedulable channels are 15–20. Additionally, six devices are configured as sniffers to monitor packets transmitted on the six channels. Then the sniffed packet with a timestamp is sent to a PC via an 8-port RS-232 PCI Express serial board.

Figure 2.5 shows our testbed. The network is deployed in a building. For each parameter configuration, 100 test cases are implemented. The generation of configurations is the same as in simulations. The configuration message is sent to devices via the gateway. Figure 2.6 shows pessimism ratios in a certain scope under different parameters. The point of the pessimism ratio 1.4 reports the number of test cases, whose pessimism ratios are between 1.4 and 1.6. When the utilization is set as 1 and the number of channels is 12, compared with PD and DM, our average

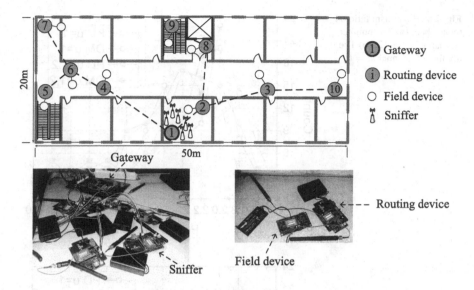

Fig. 2.5 Our testbed

pessimism ratio of 100 test cases is about 2.5 and 2.4, respectively. The result is more pessimistic than the simulations. It is because that real cases only cover a little state space. The delay observed in the real testbed is not the worst case delay, while our analysis focuses on the worst case. Therefore, for the end-to-end delay, our analysis method is more reliable than real tests.

2.5 Summary

Multiple criticality levels co-exist in real-life wireless networks. However, previous works only focus on the single-criticality network. We present an end-to-end delay analysis method for fixed priority scheduling in mixed-criticality WirelessHART networks, which can be used to determine whether all flows can be delivered to their destinations within their deadlines. In evaluations, we compare our analysis results with simulations and a testbed. The results show that the pessimism of our analysis is acceptable and reliable.

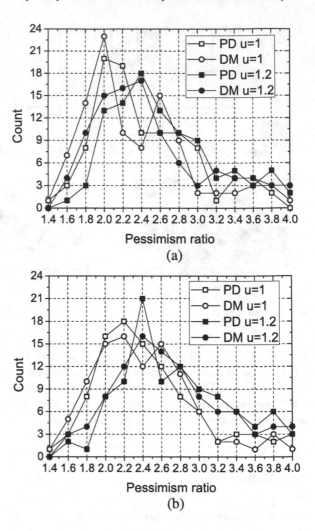

Fig. 2.6 Pessimism ratio on the testbed. (**a**) The number of channel $m = 12$. (**b**) The number of channel $m = 6$

References

1. Saifullah A, Xu Y, Lu CY, Chen YX (2011) End-to-end delay analysis for fixed priority scheduling in WirelessHART networks. In: Real-time and embedded technology and applications symposium (RTAS). IEEE, pp 13–22
2. Saifullah A, Xu Y, Lu CY, Chen YX (2010) Real-time scheduling for WirelessHART networks. In: Real-time systems symposium (RTSS). IEEE, pp 150–159
3. Saifullah A, Xu Y, Lu CY, Chen YX (2011) Priority assignment for real-time flows in WirelessHART networks. In: Euromicro conference on real-time systems (ECRTS). IEEE, pp 35–44
4. Soldati P, Zhang HB, Johansson M (2009) Deadline-constrained transmission scheduling and data evacuation in WirelessHART networks. In: The European control conference. IEEE, pp 1–7

5. Vestal S (2007) Preemptive scheduling of multi-criticality systems with varying degrees of execution time assurance. In: The IEEE real-time system symposium (RTSS). IEEE, pp 239–243

6. Niz D, Lakshmanan K, Rajkumar R (2009) On the scheduling of mixed criticality real-time task sets. In: The IEEE real time systems symposium (RTSS). IEEE, pp 291–300

7. Baruah S, Li HH, Stougie L (2010) Towards the design of certifiable mixed-criticality systems. In: The IEEE real-time and embedded technology and applications symposium. IEEE, pp 555–556

8. Baruah S, Bonifaci V, D'Angelo G, Li HH, Marchietti-Spaccamela A, Megow N, Stougie L (2012) Scheduling real-time mixed-criticality jobs. Trans Comput 61(8): 1140–1152

9. Li HH, Baruah S (2012) Global mixed-criticality scheduling on multiprocessors. In: The Euromicro conference on real-time systems (ECRTS). IEEE, pp 166–175

10. Baruah SK, Bonifaci V, D'Angelo G, Marchietti-Spaccamela A, Van Der Ster S, Stougie L (2011) Mixed-criticality scheduling of sporadic task systems. In: The 2011 European conference on algorithms. Springer, pp 555–566

11. Guan N, Ekberg P, Stigge M, Wang Y (2011) Effective and efficient scheduling of certifiable mixed-criticality sporadic task systems. In: The IEEE real-time systems symposium (RTSS). IEEE, pp 13–23

12. Huang HM, Gill C, Lu CY (2014) Implementation and evaluation of mixed criticality scheduling approaches for sporadic tasks. Trans Embed Comput Syst 13(4): 1–25

13. Tobuschat S, Axer P, Ernst R, Diemer J (2013) IDAMC: a NoC for mixed criticality systems. In: The IEEE international conference on embedded and real-time computing systems and applications. IEEE, pp 149–156

14. Diemer J, Ernst R (2010) Back suction: service guarantees for latency-sensitive on-chip networks. In: The ACM/IEEE international symposium on networks-on-chip. IEEE, pp 155–162

15. Audsley N (2013) Memory architectures for NoC-based real-time mixed criticality systems. In: The 1st workshop on mixed criticality systems. IEEE, pp 37–42

16. Burns A, Davis RI (2013) Mixed criticality on controller area network. In: The Euromicro conference on real-time systems (ECRTS). IEEE, pp 125–134

17. Herber C, Richter A, Rauchfuss H, Herkersdorf A (2013) Spatial and temporal isolation of virtual CAN controllers. In: The IEEE international conference on embedded and real-time computing systems and applications. IEEE, pp 1–7

18. Tamas-Selicean D, Pop P, Steiner W (2011) Synthesis of communication schedulers for TTEthernet-based mixed-criticality systems. In: The international conference on hardware/software codesign and system synthesis. IEEE, pp 473–482

19. Suethanuwong E (2012) Scheduling time-triggered traffic in TTEthernet systems. In: The conference on emerging technologies and factory automation. IEEE, pp 1–4

20. Steiner W (2011) Synthesis of static communication schedules for mixed-criticality systems. In: The 14th IEEE international symposium on object/component/service-oriented real-time distributed computing workshop. IEEE, pp 11–18

21. Addisu A, George L, Sciandra V, Agueh M (2013) Mixed criticality scheduling applied to JPEG2000 video streaming over wireless multimedia sensor networks. In: The 1st workshop on mixed criticality systems. IEEE, pp 1–6

22. Baruah S, Bonifaci V, D'Angelo G, Li HH, Marchietti-Spaccamela A, Megow N, Stougie L (2012) Scheduling real-time mixed-criticality jobs. Trans Electron Comput 61(8):1140–1152

23. Li HH, Baruah S (2012) Global mixed-criticality scheduling on multiprocessors. In: The 24th Euromicro conference on real-time systems. IEEE, pp 166–175

24. Guan N, Ekberg P, Stigge M, Wang Y (2011) Effective and efficient scheduling of certifiable mixed-criticality sporadic task systems. In: The 32nd IEEE real-time systems symposium. IEEE, pp 13–23

25. Joseph M, Pandya P (1986) Finding response times in a real-time system. Comput J 29(5):390–395

26. Camilo T, Silva JS, Rodrigues A, Boavida F (2007) Gensen: a topology generator for real wireless sensor networks deployment. In: The international conference on software technologies for embedded and ubiquitous systems. Springer, pp 436–445
27. Bini E, Buttazzo CC (2005) Measuring the performance of schedulability tests. Real-Time Syst 30(1):129–154
28. Liu J (2000) Real-time systems. Prentice Hall, Upper Saddle River

Chapter 3
Schedulability Analysis of Mixed-Criticality Data Under EDF Scheduling

Abstract In this chapter, to improve the schedulability of high-criticality flows when the network is running, we present a supply/demand bound function analysis method based on earliest deadline first (EDF) scheduling. In addition, our method considers both source routing and graph routing. At the beginning, when the network is in low-criticality mode, source routing is applied. When errors or exceptions occur, the network switches to high-criticality mode, and network routing turns to graph routing to guarantee that critical flows can be scheduled. By estimating the demand bound for the mixed-criticality data model, we can determine the schedulability of industrial wireless networks.

3.1 Background

Graph routing [1] as an effective way to improve network reliability has been widely used in recent years. A network under graph routing allocates two dedicated time slots for each transmission; if the first transmission fails, a retransmission will be sent. Furthermore, the controller assigns a third shared slot on a separate path for another retransmission. Since graph routing is a reliable method to handle transmission failures, a few works have begun to focus on graph routing. The work in [2] presents the first worst-case end-to-end delay analysis for periodic real-time flows under reliable graph routing. The work in [3] studies the network lifetime maximization problem under graph routing. However, graph routing introduces great challenges for real-time analysis. Many conflicts are generated on a large number of transmission tasks. Obviously, the task which is more critical but has a low priority may miss its deadline in this situation. However, many systems need to guarantee high-criticality task's schedulability even though in the worst case. That is really very important in many scenarios such as industrial production

© The Author(s) 2023
X. Jin et al., *Mixed-criticality Industrial Wireless Networks*, Wireless Networks,
https://doi.org/10.1007/978-981-19-8922-3_3

line, vehicle driving system, etc. To improve the schedulability of high-criticality flows when the network is running, we introduce resource analysis into mixed-criticality industrial wireless networks. Mixed-criticality network can improve the schedulability of high-criticality flows by dynamically switching the network criticality, and resource analysis is a major way to analyze the schedulability in real-time systems. Combining mixed-criticality network and resource analysis, we can estimate the schedulability of networks with different critical levels.

In this chapter, we propose a novel industrial network model with EDF schedul-ing. Our objective is to improve the network reliability, especially for high-criticality flows to arrive at their destinations on time even though in the worst case. We analyze the network schedulability by the method of resource analysis. The network is reliable when the network resource supply is no less than the network upper demand in any length of time slot. The main challenges in our work are (1) how to evaluate network demand when a network switches from low-criticality mode to high-criticality mode and (2) how to tighten the network demand bound function to ensure that the analysis result is not too pessimistic. The network we focus on, in the beginning, works in low-criticality mode, and the flows transmit under the EDF policy [4] and source routing. The packets are transmitted from the source to the destination on the primary paths; when an error occurs or the demand changes, the network switches to high-criticality mode to enhance the schedulability of high-criticality flows. The network substitutes reliable graph routing for source routing. Furthermore, we present a supply/demand bound analysis method to analyze the schedulability of periodic flows in industrial wireless sensor networks. By comparing the relationship between network supply bound and demand, we can predict whether the network can be scheduled. The current study makes the following key contributions:

1. We propose a mixed-criticality industrial network, in which network routing switches from source routing to graph routing when the criticality mode changes.
2. We theoretically derive the supply/demand bound function as a novel analysis method for industrial wireless networks. By analyzing channel contention and transmission conflict, we obtain the upper-bound function of demand in any length of time slot. When given a network supply bound function, we can determine the schedulability of flows under different criticality modes.
3. We tighten the demand bound by analyzing *carry-over jobs* (which are released but not finished at the switching slot) and discussing the number of conflicts between two flows.
4. Our method can be applied for general networks. By calculating the maximum demand bound of networks, we can analyze network schedulability in the system design stage; after network deployment, the upper bound of communications can be obtained by our method.

3.2 System Model

We consider an industrial wireless network consisting of field devices, one gateway, and one centralized network manager. Our network is proposed in three aspects. We first propose a network model that is abstracted from mainstream industrial network standards. Then, we introduce a mixed-criticality network. Finally, we apply EDF scheduling in the industrial network.

3.2.1 Network Model

Without loss of generality, our model has the same salient features as WirelessHART and WIA-PA, which make it particularly suitable for process industries:

Time Division Multiple Access (TDMA) In industrial wireless sensor networks, time is synchronized and slotted. Because the length of a time slot allows exactly one transmission, TDMA protocols can provide predictable communication latency and real-time communication.

Route and Spectrum Diversity To mitigate physical obstacles, broken links, and interference, the messages are routed through multiple paths. Spectrum diversity gives the network access to all 16 channels defined in the IEEE 802.15.4 physical layer and allows per-time slot channel hopping. The combination of spectrum and route diversity allows a packet to be transmitted multiple times, over different channels and different paths, thereby handling the challenges of network dynamics in harsh and variable environments at the cost of redundant transmissions and scheduling complexity [5].

Handling Internal Interference Industrial networks allow only one transmission in each channel in a time slot across the entire network, thereby avoiding the spatial reuse of channels. Thus, the total number of concurrent transmissions in the entire network at any slot is no greater than the number of available channels.

With the above features, the network can be modeled as a graph $G = (V, E, m)$, in which the node set V represents the network devices (all sensor nodes in our model are fixed), E is the set of edges between these devices, and m is the number of channels. Network routing is shown in Fig. 3.1; our model supports both source routing and graph routing. Source outing is well known in academic research; we will not explore it in this article. Graph routing is a unique feature of industrial

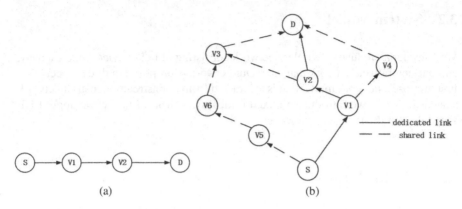

Fig. 3.1 Network routing. (**a**) Source routing. (**b**) Graph routing

wireless sensor networks. In graph routing, a routing graph is a directed list of paths that connect two devices. As shown in Fig. 3.1b, graph routing has a primary path and multiple backup paths. This provides redundancy in the route and improves the reliability. As stated in the standard of WirelessHART, for each intermediate node on the primary path, a backup path is generated to handle link or node failure on the primary path. The network manager allocates α dedicated slots, a transmission and $(\alpha - 1)$ retransmission on the primary path. A $(\alpha + 1)$th shared slot is allocated on the backup path, usually $\alpha = 2$. In a dedicated slot, one channel only allows one transmission. However, for the case of a shared slot, the transmissions having the same receiver can be scheduled in the same slot. The senders that attempt to transmit in a shared slot contend for the channel using a carrier sense multiple access with collision avoidance (CSMA/CA) scheme [2]. Hence, multiple transmissions can be scheduled in the same channel to contend in a shared slot. For instance, the network manager allocates two dedicated slots for the packet transmits from node S to node V_1 in Fig. 3.1b. After the transmissions on the primary path, a third slot is allocated for the packet transmits from node S to V_5 as a backup path. When two backup paths intersect at node V_3, they can avoid collision by CSMA/CA.

It is important to note that the receiver responds with an ACK packet before retransmission and backup; the sender retransmits or sends a backup packet when it does not receive an ACK. Because ACK is a part of the transmission, we do not need to especially analyze the demand of ACK.

3.2.2　Mixed-Criticality Network

A periodic end-to-end communication between a source and a destination is called a *flow*. Network switch instruction is a part of the control flow. Because we analyze network total demand, we need not distinguish whether a flow is a data flow or a control flow. The total number of flows in the network is n, denoted by $F = \{F_1, F_2, \ldots, F_n\}$. F_i is characterized by $< t_i, d_i, \xi, c_i, \phi_i >, 1 \leq i \leq n$, where t_i is the period; d_i is the deadline; ξ is the criticality level (we focus on dual-criticality network $\{LO, HI\}$); $\xi = 2$, means the network allocates two slots, one transmission and one retransmission. Our model can be easily extended to networks with an arbitrary number of criticality levels (by increasing the number of retransmissions on the primary path); c_i is the number of hops required to deliver a packet from source to destination. When the network mode switches to high criticality, we denote the total transmission hops of both the primary path and shared paths as C_i; and ϕ_i is the routing path of the flow. Thus, we can describe each flow F_i as follows. F_i periodically generates a packet at its period t_i, and then sends it to the destination before its deadline d_i via the routing path ϕ_i with c_i hops.

In the beginning, messages are transmitted under source routing in low criticality. When an error occurs or the demand changes, the control flow will send a switch instruction, and the network will switch to high-criticality mode. To enhance network reliability, the messages are transmitted under graph routing when the network is running on high-criticality mode. This is an irreversible process; high-criticality mode will never switch back to low-criticality mode (the analytical method of irreversible processes is similar to criticality mode switches from low to high). After the switch, we are not required to meet any deadlines for low-criticality flows, but high-criticality flows may instead execute for up to their high-criticality level characters.

3.2.3 EDF Scheduling in Industrial Networks

In this subsection, we provide an overview of the earliest deadline first scheduling under industrial wireless sensor networks to analyze network schedulability. EDF scheduling is a commonly adopted policy in practice for real-time CPU scheduling, cyber-physical systems, and industrial networks [6]. In an EDF scheduling policy, each job priority is assigned by its absolute deadline, and the transmission is scheduled based on this priority. Each node in our network is equipped with a half-duplex omnidirectional radio transceiver that can alternate its status between transmitting and receiving. There are two kinds of delay in industrial wireless sensor networks, which can be summarized as follows:

- Channel contention: each channel is assigned to one transmission across the entire network in the same slot.
- Transmission conflicts: whenever two transmissions conflict, the transmission that belongs to the lower-priority job must be delayed by the higher-priority one, regardless of how many channels are available. It is important to note that one node can perform only one operation (receiving or transmitting) in each slot.

In EDF scheduling, the priority is inversely proportional to its absolute deadline. We explain the operating principle of EDF scheduling in Fig. 3.2. There are two

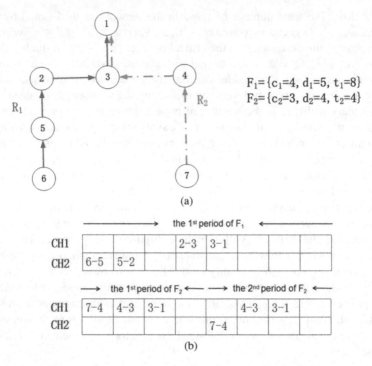

Fig. 3.2 An example for EDF scheduling. (a) Routing. (b) EDF scheduling

channels(CH1 and CH2) and flows in this network. At the beginning, the priority of F_2 is higher than F_1 since $d_2 = 4 < d_1 = 5$. Then the controller allocates CH1 for F_2 first. The flow with lower priority must be delayed when transmission conflict occurs such as F_1 will be delayed by F_2 at the 3rd time slot. At the 5th time slot, the second packet is generated by F_2 with an absolute deadline 8, which is larger than 5. Hence, the priority inversion, and CH1 are allocated to F_1.

Channel contention occurs when high-priority jobs occupy all channels in a time slot; a transmission conflict is generated when several transmissions involve a common node at the same dedicated slot, and a low-priority job is delayed by high-priority ones. However, for the case of shared slots, transmissions with the same receiver can be scheduled in the same slot. When channel contention occurs between backup paths, the senders on the backup path use a CSMA/CA scheme to contend for the channel, and a network delay will not result in this condition. For a network under graph routing, two paths ϕ_i and ϕ_j involving a common node may conflict in four conditions:

1. ϕ_i is a primary path, ϕ_j is a backup path;
2. both ϕ_i and ϕ_j are primary paths;
3. both ϕ_i and ϕ_j are backup paths;
4. ϕ_i is a backup path, ϕ_j is a primary path.

Except for condition 3, the other three conditions may generate transmission conflicts. Consequently, the total delay caused by these conditions depends on how their primary and backup paths intersect in the network.

In a real-time system, one task is schedulable when it could be executed completely before its deadline. Hence, the flow could be scheduled when all the packets generated by the flow could arrive destination before their relative deadlines. Then we define the network schedulability as whether or not all flows in a network are schedulable.

3.3 Problem Formulation

Given a mixed-criticality industrial network $G = (V, E, m)$, the flow set F and the EDF scheduling algorithm, our objective is to analyze the relationship between the maximum execution demand of the flows and network resource in any time interval such that the schedulability of the flow set can be determined. A successful method to analyzing the schedulability of real-time workloads is to use demand bound functions [7, 8]. We introduce this concept into industrial wireless sensor networks and propose two definitions as follows:

Definition 3.1 (Supply Bound Function) A supply bound function sbf(l) is the minimal transmission capacity provided by the network within a time interval of length l.

Definition 3.2 (Demand-Bound Function) A demand bound function dbf(F_i, l) gives an upper bound on the maximum possible execution demand of flow F_i in any time interval of length l, where demand is calculated as the total amount of required execution time of flows with their whole scheduling windows within the time interval.

There are methods for computing the supply bound function $sbf(l)$ in single-processor systems [9, 10]—for example, a unit-speed, dedicated uniprocessor has $sbf(l) = l$. We say that a supply bound function sbf is of no more than unit speed if

$$sbf(0) = 0 \wedge \forall l, k \geq 0 : sbf(l + k) - sbf(l) \leq k. \qquad (3.1)$$

Because each channel can be mapped as one processor, the supply bound function sbf of the industrial network can be bounded as

$$sbf(0) = 0 \wedge \forall l, k \geq 0 : sbf(l + k) - sbf(l) \leq Ch \times k, \qquad (3.2)$$

where Ch is the number of channels in the network. Furthermore, as a natural assumption of all proposed virtual resource platforms in the literature, we assume that the supply bound function is piecewise linear in all intervals $[k, k + l]$. In TDM (time division multiple), the network supply bound function can be expressed as

$$sbf(l) = \max(l \mod \Theta - \Theta + \Phi, 0) + \lfloor \frac{l}{\Theta} \rfloor \Phi, \tag{3.3}$$

where Θ is the period of TDM, and Φ is the length of slots allocated to the transmission.

In different modes, the schedulability of the flow set is determined as follows:

$$\sum_{F_i \in F} dbf_{LO}(F_i, l) \leq sbf_{LO}(l), \forall l \geq 0. \tag{3.4}$$

$$\sum_{F_i \in HI(F)} dbf_{HI}(F_i, l) \leq sbf_{HI}(l), \forall l \geq 0. \tag{3.5}$$

Similar to real-time scheduling, the flow set is scheduled when the network is satisfied by Eqs. (3.4) and (3.5). However, in contrast to real-time scheduling, there are two kinds of delays in industrial wireless sensor networks, channel contention and transmission conflicts. When a transmission conflict occurs, a high-priority job will influence a low priority job, and thus, the flows are not independent.

Note that transmission conflict is a distinguishing feature in industrial wireless sensor networks that does not exist in conventional real-time processor scheduling problems. To analyze the network demand in any time interval, we must consider the delay caused by transmission conflicts.

Moreover, in mixed-criticality networks, there may be some jobs that are released but not finished at the time of the switch to high-criticality mode; we define these jobs as *carry-over jobs*. We must analyze *carry-over jobs* to tighten the demand bound of the network.

3.4   Demand-Bound Function of Industrial Networks

In this section, we analyze the network demand bound function for a single-criticality network and mixed-criticality network. For the single-criticality network, we study the demand bound function from channel contention and transmission conflicts. On this basis, we then analyze the delay caused by *carry-over jobs* (the job that is released but not finished at the time of the switch) in the mixed-criticality network. Finally, we study the methods for tightening the network demand bound function.

3.4.1 Analysis of Single-Criticality Networks

In this subsection, we study the demand bound function under a single-criticality network in two steps. First, we formulate network transmission conflict delay with path overlaps; we then analyze the network dbf. To make our study self-contained, we present the results of the state-of-the-art demand bound function for CPU scheduling [11, 12]. Assuming that the flows are executed on a multiprocessor platform, the channel is mapped as a processor. We can obtain the network demand caused by channel contention in any time interval l as

$$dbf(l)^{ch} = \frac{1}{m} \sum_{i=1}^{n} \left[\left(\left\lfloor \frac{l - d_i}{t_i} \right\rfloor + 1 \right) c_i \right]_0. \tag{3.6}$$

Equation (3.6) considers only the delay caused by channel contention, denoted as $dbf(l)^{ch}$. The jobs are conflicted when their transmission paths have overlaps. As shown in Fig. 3.3, the priority of the job in F_i is higher than the one in F_j, so the job in F_j may be delayed by in F_i at nodes V and V_1 to V_h (we assume the network is connected and do not consider the case where the path disconnects).

Transmission conflicts are generated at the path overlaps, and the network requires more resources to solve the transmission conflicts. To obtain $dbf(l)$ of the network, we must first study the relationship between conflict delay and path overlap. However, estimation transmission conflict delay by the length of the overlap is often a pessimistic method. As shown in Fig. 3.3, the delay is much smaller than the length of the path overlap. To avoid pessimistic estimation, we introduce the result proposed by Saifullah in [13]. The length of the kth path overlap is denoted as $Len_k(ij)$, and its conflict delay is $D_k(ij)$. For the overlap as $V_1 \rightarrow \ldots V_h$, if there exists node $u, w \in V$ such that $u \rightarrow V_1 \rightarrow \ldots V_h \rightarrow w$ is also on F_i's route, then $Len_k(ij) = h + 1$. If only u or only w exists, then $Len_k(ij) = h$. If neither u nor v exists, then $Len_k(ij) = h - 1$. In our example, $Len_1(ij) = 2$, $Len_2(ij) = 7$ and $D(ij) = D_1(ij) + D_2(ij)$, which is at most $2 + 3 = 5$. Obviously, $Len_k(ij)$ is the upper bound of $D_k(ij)$, which means $Len_k(ij) \geq D_k(ij)$. For the flow set F, the total delay caused by transmission conflicts Δ is

$$\Delta = \sum_{1 \leq i, j \leq n} D_k(ij) \leq \sum_{1 \leq i, j \leq n} Len_k(ij). \tag{3.7}$$

By the Lemma proposed in [13], the estimation of the delay caused by overlap with a length of at least 4 can be tightened. We then formulate the total transmission conflicts between F_i and F_j as

$$\Delta(ij) = \sum_{k=1}^{\delta(ij)} Len_k(ij) - \sum_{k'=1}^{\delta'(ij)} (Len_{k'}(ij) - 3), \tag{3.8}$$

Fig. 3.3 An example of
transmission delay

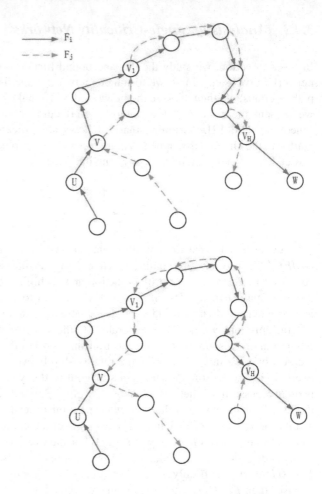

where $\delta(ij)$ is the number of path overlaps, $\delta'(ij)$ is the number of paths overlap
with a length of at least 4. Because all flows have a periodic duty, we denote T as the
least common multiple of flow set F (because the period is an integral multiple of 2,
T is equal to the longest period among F). Network dbf changes with time interval l
while it slides from 0 to T. However, Lemma 3.2 proposed by Saifullah is scheduled
under fixed priority, so the priorities of flows are variable under EDF scheduling.
We must analyze whether Saifullah's result is suitable under EDF scheduling. We
denote the mth job generated by F_i as F_i^m, and our objective is to estimate the delay
caused by transmission conflicts by analyzing the number of conflicts.

Lemma 3.1 F_i^k *and* F_j^g *are two jobs of flow i and j, when* F_i^k *and* F_j^g *(* $F_i^k \in$
$hp(F_j^g)$ *) conflict, the job* F_i^k *will never be blocked by the job* F_j^{g+m}. *However,*
F_i^{k+m} *may be blocked by* F_j^g.

Proof At the beginning, the priority of F_i^k is higher than F_j^g, which means $d_i^k < d_j^k$. As Fig. 3.3 shows, two flows may conflict at V_1, and F_j is delayed by F_i. When F_i^k is forwarded to V_h, two jobs may conflict again. If F_i^k is blocked by F_j^{k+m}, we can obtain $d_i^k > d_j^{g+m}$. Because $d_j^{g+m} > d_j^g$, this contradicts with $d_i^k < d_j^g$. Hence, F_i^k will never be blocked by F_j^{g+m}.

We prove that F_i^{k+m} is blocked by F_j^g through an example. We use the following simple flow set: $F_1 = \{c_1 = 1, d_1 = t_1 = 2\}$ and $F_2 = \{c_2 = 1, d_2 = t_2 = 3\}$.

At the beginning, the priority of F_1^1 is higher than F_2^1, because the absolute deadline is 2 and 3, respectively. At time slot 2, another job is generated by F_1 with the absolute deadline of 2. However, the absolute deadline of F_2^1 is 1, F_1^2 is blocked by F_2^1. Hence, F_i^{k+m} can be blocked by F_j^g.   □

Because a path is a chain of transmissions from source to destination, in considering the conflict delay caused by multiple jobs of F_i on flow F_j, we analyze the number of conflicts for F_i and F_j. Thus, Lemma 3.2 establishes the upper bound of this value.

Lemma 3.2 *When F_j and F_i conflict, within any time interval of length l, each job of F_j can be blocked no more than $\lceil \frac{l}{t_i} \rceil$ times, and F_j can be blocked by F_j no more than $\lceil \frac{l}{t_j} \rceil$ times.*

Proof Based on Lemma 3.1, we know that the priority inversion will occur in the process of transmission. If F_i^k is a higher-priority job than F_j^g, the jobs released after F_j^g must be blocked by F_i^k until F_i^k is finished. If all jobs generated by F_i satisfy $d_i^{k+\lceil \frac{l}{t_i} \rceil} < d_j^g$, where k and g are the first jobs for F_i and F_j, respectively, in l, then there are no more than $\lceil \frac{l}{t_i} \rceil$ jobs of F_i. Beyond that, because there is no transmission conflict, the other jobs of F_j are not blocked by F_i. Hence, F_j can be blocked by F_i no more than $\lceil \frac{l}{t_i} \rceil$ times. The same as F_i, F_i can be blocked by F_j no more than $\lceil \frac{l}{t_j} \rceil$ times.

□

By Lemmas 3.1 and 3.2, we can estimate the network demand caused by the transmission conflict. Based on Eq. (3.6), we obtain the upper bound of $dbf(l)$ as follows:

Theorem 3.1 *In any time interval of length l, the demand bound function under a single-critical network (low-criticality mode) is upper-bounded by*

$$dbf_{LO}(l) = \frac{1}{m} \sum_{i=1}^{n} \left[\left(\left\lfloor \frac{l-d_i}{t_i} \right\rfloor + 1 \right) c_i \right]_0 + \sum_{1 \le i,j \le n} (\Delta(ij) \max\{\lceil \frac{l}{t_i} \rceil, \lceil \frac{l}{t_j} \rceil\}).$$

$$(3.9)$$

Proof Network demand is the upper bound in a time interval of length l, which consists of two parts, channel contention and transmission conflict. The demand of channel contention is bounded by Eq. (3.6). For the demand of the transmission conflict, we first analyze each time conflict delay for every two paths by Eq. (3.8); the number of conflicts can then be obtained by Lemma 3.2. We can obtain the network demand of transmission conflict as

$$\sum_{1 \le i, j \le n} (\Delta(ij) \max\{\lceil \frac{l}{t_i} \rceil, \lceil \frac{l}{t_j} \rceil\}). \tag{3.10}$$

Hence, we can obtain the demand bound function under a single-critical network upper-bounded by Eq. (3.9).

□

3.4.2 Analysis of Mixed-Criticality Networks

Based on the result proposed in Sect. 3.4.1, we extend the idea of a demand bound function to a mixed-criticality network. For illustration purposes, only a dual-criticality network is considered; this means that ξ has only two values, LO (low-criticality mode) and HI (high-criticality mode). Nevertheless, it can be easily extended to networks with an arbitrary number of criticality modes. We construct three demand bound functions: the demand bound function in low- and high-criticality modes ($dbf_{LO}(l)$ and $dbf_{HI}(l)$) and the demand bound function when network mode switches ($dbf_{LO2HI}(l)$). We analyze $dbf_{HI}(l)$ and $dbf_{LO2HI}(l)$ under graph routing in this subsection.

The network begins from the low-criticality level, and all flows are served and executed as in a single-criticality network. When errors or emergencies occur, the centralized network manager will trigger the switching of the network mode from LO to HI. In high-criticality mode, the network turns to graph routing, and the flows in the low-criticality level are discarded; only high-criticality flows can be delivered. The job that is active (released, but not finished) from a high-criticality flow at the time of the switch is still running under source routing; n_{HI} is the number of high-criticality flows, and there are no more than n_{HI} *carry-over jobs* that are active at the time of the switch. We define these *carry-over jobs* as new flows $F_{(n_{HI}+1)}, F_{(n_{HI}+2)} \ldots F_{2n_{HI}}$, which have the same characters as the corresponding flows in F except for c and t. For the new flow $F_{p+n_{HI}}$, $c_p > c_{(p+n_{HI})}$, and as an accidental event, $t_{(p+n_{HI})} \gg t_p$.

When the network switches from LO to HI, the demand of *carry-over jobs* is

$$\frac{1}{m} \sum_{p=1+n_{HI}}^{2n_{HI}} c_p + \sum_{n_{HI} \le p, q \le 2n_{HI}} \Delta(pq). \tag{3.11}$$

Furthermore, the flows will generate new jobs when the network switches to high-criticality mode. Because each node except the destination on the primary path generates one backup path, the total number of paths for F_p is $c_p + 1$ and the execution time for each backup path c_p^k can be obtained from the network easily. The total execution time of F_i can be denoted as $C_p = c_p + \sum_{k=1}^{c_p} c_p^k$. Therefore, network demand for channel contention under graph routing is

$$dbf_{HI}^{ch}(l) = \frac{2}{m} \sum_{p=1}^{n_{HI}} \left[\left(\left\lfloor \frac{l - d_p}{t_p} \right\rfloor + 1 \right) C_p \right]_0. \qquad (3.12)$$

Based on the rules of transmission conflict proposed in Sect. 3.2.3, a transmission conflict between two flows is generated only if there is at least one flow transmission on the primary path. Therefore, we analyze $dbf_{HI}(l)$ by studying the transmission conflict generated on the primary path. For F_p^g and F_q^m, when given $d_p < dq$, F_q^m may be delayed by F_p^g and its backup paths. We denote the path set of F_p and its backup paths as I; each path in I is denoted as p'. The upper bound delay of F_q^m caused by F_p^g is denoted as $\Delta(p'q)$. $\Delta(p'q)$ can be formulated as

$$\Delta(p'q) = \sum_{p'=1}^{c_p+1} \left(\sum_{k=1}^{\delta(p'q)} Len_k(p'q) - \sum_{k'=1}^{\delta'(p'q)} (Len_{k'}(p'q) - 3) \right). \qquad (3.13)$$

For the job on the backup path, a transmission delay occurs only when it conflicts with primary paths with high-priority jobs. When we reverse the priority of F_p^g and F_q^m, Eq. (3.13) is the upper bound additional demand of F_p^g caused by F_q^m. From the above, the network upper bound demand function under graph routing can be described as

$$dbf_{HI}(l) = \frac{2}{m} \sum_{i=1}^{n_{HI}} \left[\left(\left\lfloor \frac{l - d_i}{t_i} \right\rfloor + 1 \right) C_i \right]_0$$

$$+ \sum_{1 \le p,q \le n_{HI}} (\Delta(p'q) \max\{\lceil \frac{l}{t_p} \rceil, \lceil \frac{l}{t_q} \rceil\}). \qquad (3.14)$$

We can then obtain $dbf_{LO2HI}(l)$ as

$$dbf_{LO2HI}(l) = \frac{2}{m} \sum_{p=1}^{n_{HI}} \left(\left[\left(\left\lfloor \frac{l - d_p}{t_p} \right\rfloor + 1 \right) C_p + \frac{1}{2} c_p \right) \right]_0$$

$$+ \sum_{1 \le p,q \le 2n_{HI}} (\Delta(p'q) \max\{\lceil \frac{l}{t_p} \rceil, \lceil \frac{l}{t_q} \rceil\}). \qquad (3.15)$$

Because transmission on a backup path occurs only when the two previous attempts fail, when the transmission success rate on the primary path satisfies the network packet reception ratio, the sender has no need to send a backup packet. Hence, the network upper bound demand function in this case can be rewritten as

$$dbf_{LO2HI}(l) = \frac{3}{m} \sum_{p=1}^{n_{HI}} \left[\left(\lfloor \frac{l - d_p}{t_p} + 1 \rfloor \right) c_p \right]_0$$

$$+ \sum_{1 \le p,q \le 2n_{HI}} (\Delta(pq) \max\{\lceil \frac{l}{t_p} \rceil, \lceil \frac{l}{t_q} \rceil\}). \qquad (3.16)$$

3.4.3 Tightening the Demand Bound Functions

A loose demand bound function will lead to a pessimistic estimation of network schedulability. In this subsection, we tighten our demand bound functions by discussing the relationship between two flows and transmission conflict.

In our previous analysis Lemma 3.2, the number of conflict jobs is a conservative estimation as $\max\{\lceil \frac{l}{t_i} \rceil, \lceil \frac{l}{t_j} \rceil\}$. However, this value can be reduced by classifying discussions. We divide this value into the following categories:

- $t_i \le t_j$, and $d_i \le d_j$.
- $t_i \le t_j$, and $d_i \ge d_j$.

When the path of F_i and F_j have overlaps, they may generate transmission conflicts. The delay caused by conflict cannot occur in each slot because the flow does not transmit between d and t. Obviously, when one flow works in its ideal time (between d and t), there is no transmission conflict between F_i and F_j.

Condition 1 is shown in Fig. 3.4a; conflict occurs only when both F_i and F_j have job transmissions on the path. For a given l, the number of conflicting jobs can be expressed as

$$\lceil \frac{l}{t_j} \rceil (\lfloor \frac{d_j}{t_i} \rfloor + 1). \qquad (3.17)$$

Similarly, we can obtain the number of conflicting jobs in condition 2 as

$$\lceil \frac{l}{t_j} \rceil (\lfloor \frac{d_j}{t_i} \rfloor + 1) = 2\lceil \frac{l}{t_j} \rceil. \qquad (3.18)$$

We denote the number of conflicts as $Num(ij)$. When we know each flow's routing information, the estimation of $Num(ij)$ can be further precise. By taking the modulus of $\frac{d_j}{t_i}$, we can estimate the maximum length of F_i's residual path as

Fig. 3.4 Classified discussion. (**a**) Condition 1. (**b**) Condition 2

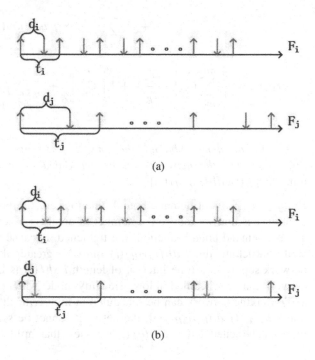

$||\frac{d_j}{t_i}||$. The delay on this residual path is denoted as ψ, and we can obtain ψ as follows:

- If F_i has an overlap with F_j on this residual path, $\psi = \Delta(||\frac{d_j}{t_i}||)$, where $\Delta(||\frac{d_j}{t_i}||)$ is the delay on the residual path whose length is $||\frac{d_j}{t_i}||$.
- If F_i has no overlap with F_j on this residual path, $\psi = 0$.

The number of conflicts can be expressed as

$$Num(ij) = \lceil \frac{l}{t_j} \rceil (\lfloor \frac{d_j}{t_i} \rfloor + \psi). \tag{3.19}$$

We can then obtain the network demand bound functions as follows:

Theorem 3.2 *In any time interval of length l, the demand bound function in each mode can be expressed as*

$$dbf_{LO}(l) = \frac{1}{m} \sum_{i=1}^{n} \left[\left(\left\lfloor \frac{l - d_i}{t_i} \right\rfloor + 1 \right) c_i \right]_0 + \sum_{1 \le i, j \le n} (\Delta(ij)Num(ij)). \tag{3.20}$$

$$dbf_{LO2HI}(l) = \frac{2}{m} \sum_{p=1}^{n_{HI}} \left(\left[\left(\left\lfloor \frac{l - d_p}{t_p} \right\rfloor + 1 \right) C_p + \frac{1}{2} c_p \right]_0 \right.$$

$$+ \sum_{1 \le p,q \le 2n_{HI}} (\Delta(p'q)Num(pq)). \tag{3.21}$$

$$dbf_{HI}(l) = \frac{2}{m} \sum_{p=1}^{n_{HI}} \left(\left[\left(\left\lfloor \frac{l-d_p}{t_p} + 1 \right\rfloor \right) C_p \right) \right]_0 + \sum_{1 \le p,q \le 2n_{HI}} (\Delta(p'q)Num(pq)).$$

$$\tag{3.22}$$

The network demand bound function is $dbf(l) = \max\{dbf_{LO}(l), dbf_{LO2HI}(l), dbf_{HI}(l)\}$, and the network can be scheduled when $dbf(l)$ is no less than $\min\{dbf_{LO}(l), dbf_{LO2HI}(l)\}$.

Proof The proofs of demand bound functions are similar to in Theorem 3.1. The difference is that we reduce the number of conflicts by classifying the discussion, and the demand bound functions are tightened. Because there are *carry-over jobs* at the switching time, $dbf_{LO2HI}(l)$ must be greater than $dbf_{HI}(l)$. When the network supply in a time interval of length l $sbf(l)$ is larger than $dbf_{LO}(l)$, the network can be scheduled in low-criticality mode; when $dbf_{LO2HI}(l) \le sbf(l) < dbf_{LO}(l)$, the network can be scheduled in high-criticality mode; when $sbf(l) > \max\{dbf_{LO}(l), dbf_{LO2HI}(l)\}$, the network cannot be scheduled. Hence, the network can be scheduled when $dbf(l)$ is no less than $\min\{dbf_{LO}(l), dbf_{LO2HI}(l)\}$.

□

3.5   Performance Evaluations

In this section, we conduct experiments to evaluate the performance of our proposed methods. Our method is first compared with the simulation result. We then compare our method with the supply/demand bound function analysis without tightening.

To illustrate the applicability of our method, for each parameter configuration, several test cases are generated randomly. For each test case, the network gateway is placed at the center of playground area A, and the other nodes are deployed randomly around the gateway. According to the suggestion in [14], given the transmitting range $d = 40$ m, the number of nodes n and the playground area A should satisfy

$$\frac{n}{A} = \frac{2\pi}{d^2\sqrt{27}}. \tag{3.23}$$

If two nodes can communicate with each other, which means that the distance between two nodes is less than d, they are adjacent nodes. By repeatedly connecting the nearest node from the source node to the gateway, the network topology can be obtained. If some source nodes cannot connect to the gateway, their locations are generated randomly again.

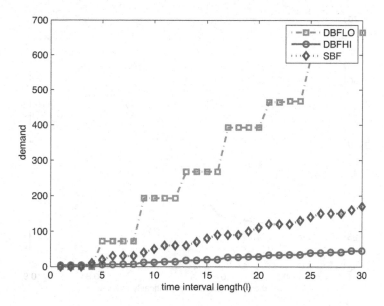

Fig. 3.5 Relationship between demand bound functions and supply bound function

Our simulations use the utilization u to control the workload of the entire network. To make flow sets available, we specify the network utilization $U = \sum u_i (U < 1)$, and the UUniFast algorithm [15] is used to generate each flow's utilization u_i ($u_i = \frac{c_i}{t_i}$). The result generated by the UUniFast algorithm follows a uniform distribution and is neither pessimistic nor optimistic for the analysis [15].

Figure 3.5 is an example of the relationship between the demand bound function in different criticality modes and the supply bound function. In this example, according to the actual situation, we set the number of nodes as $n = 70$ and the number of flows as $F = 20$. At the beginning, with the network running in low-criticality mode, the demand is zero. At time slot 5, $DBFLO$ is 72, which is larger than the upper bound of network supply; the network then switches to high-criticality mode. Considering *carry-over jobs*, we can calculate the demand in high-criticality mode from time slot 5. Because the network demand is less than the supply, this example is a stable network. Furthermore, Fig. 3.5 reveals that the demand bound functions are stepwise increasing. This is because $dbf(l)$ is the network demand over a period of time. When a job has enough time slots to transmit (e.g., a job is just released), its demand is zero and does not require immediate execution. With the decrease of the remaining time, the job becomes urgent. When the remaining time for the job is c, the job must be forwarded immediately; otherwise, it will miss the deadline. The job demand is then changed to the number of hops c.

Figure 3.6 is the variation tendency of $DBFHI$ with the proportion of high-criticality flows. Because changing the proportion of n_{HI} does not affect network demand in low-criticality mode, Fig. 3.6 shows the network demand only in high-criticality mode. Obviously, the network demand is increasing with the increasing

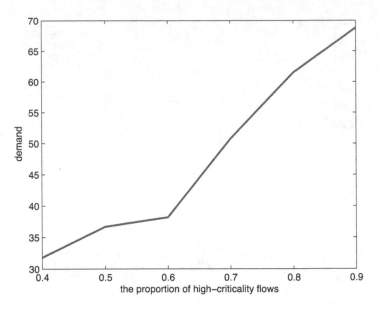

Fig. 3.6 Variation tendency of $DBFHI$ with the percentage of high-criticality flows

proportion of high-criticality flows. At the beginning (0.4–0.6), the network demand increases slowly. From 0.7–0.9, the demand of the network increases rapidly. This is because more flows in high-criticality mode generate more transmission conflicts in conditions 1, 2, and 4. The network needs more resources to ensure that the job meets its deadline. This phenomenon is enhanced severely with increasing P.

To analyze the correctness of our method, we compare the network schedulability ratio between the simulation result (denoted as MixedSim) and our method (denoted as MixedEDF) in Fig. 3.7. For each point in the figures, more than 100 test cases are randomly generated. From the figures, we can know that our algorithm can accurately evaluate the network schedulability ratio regardless of which parameters are used. Because we pessimistically estimate transmission conflicts to guarantee our method's reliability, the evaluation value of the network demand bound is larger than the actual demand. In Fig. 3.7a and b, the proportions of high-criticality flows are $P = 0.4$ and $P = 0.5$, respectively. With the increasing of nodes, the network schedulability ratio declines in both situations. However, the schedulability ratio in Fig. 3.7b falls faster than in Fig .3.7a. This is because the network generates more transmission conflicts when increasing the number of high-criticality flows. Note that compared with Fig. 3.7a, Fig. 3.7c has 0.1 additional utilization, so the spacing between the simulation curve and analysis curve is expanded. Although there are fluctuations between 30 to 60, our method can always bound the schedulable ratio (the fluctuations are caused by the randomly generated network environment). Because the two figures generate test cases according to the respective utilization, their test cases are different. When network utilization increases, the number of

Fig. 3.7 Relationship
between schedulability ratio
and the number of nodes. (**a**)
$U = 0.5$, $P = 0.4$. (**b**)
$U = 0.5$, $P = 0.5$. (**c**)
$U = 0.6$, $P = 0.4$

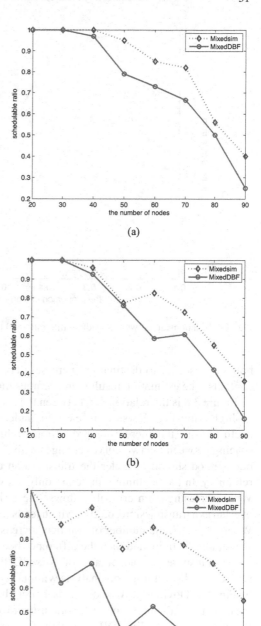

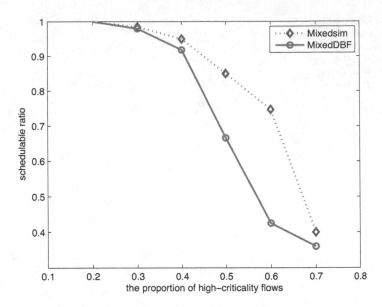

Fig. 3.8 Relationship between schedulability ratio and the proportion of high-criticality flows

hops from source to destination increases. This increases the number of potential conflicts. The estimation result then becomes more pessimistic.

Figure 3.8 is the relationship between the schedulability ratio and the proportion of high-criticality flows. It is easy to understand that the schedulability ratio declines with the increasing proportion of high-criticality flows. However, the spacing between the two curves changes with P (small–big–small). This is because our method should consider the transmission conflicts in all situations to ensure reliability. In the beginning, there are only a few conflicts in high-criticality mode. With increasing high-criticality flows, the strict estimation considers each path overlap as a transmission conflict, which leads to larger spacing between two curves. When $P = 0.7$, the number of conflicts increases in MixedSim, which reduces the schedulability ratio, and then the difference becomes small.

We illustrate the advantage of MixedDBF by comparing it with the supply/demand bound function analysis without tightening (denoted as MixedDBF-nt) in Fig. 3.9. Obviously, MixedDBF is better than MixedDBF-nt regardless of the conditions. With increasing network utilization or proportion of high-criticality flows, the error of MixedDBF-nt grows faster than MixedDBF. The reason is that both increasing network utilization and the number of high-criticality flows will increase the number of path overlaps. MixedDBF tightens the delay caused by the transmission conflict by Eq. (3.19). With increasing overlaps, the effect of Eq. (3.19) will be better. Hence, the error of MixedDBF-nt grows faster than MixedDBF.

Fig. 3.9 Schedulability
comparison among
MixedSim, MixedDBF,
MixedDBF-nt. (**a**)
$U = 0.4$, $P = 0.2$. (**b**)
$U = 0.5$, $P = 0.2$. (**c**)
$U = 0.4$, $P = 0.6$

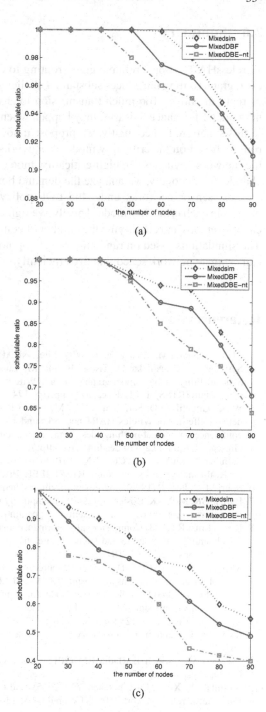

3.6 Summary

WirelessHART adopts reliable graph routing to enhance network reliability. However, graph routing introduces substantial challenges in analyzing the schedulability of real-time flows. Too much transmission load will increase conflicts and reduce network performance. Disaster may happen when critical tasks miss their deadlines in this situation. Hence, firstly, we propose a novel network model that can switch routing based on the criticality mode of networks. When errors or accidents occur, the network switches to high-criticality mode and low-level critical tasks are abandoned. Secondly, we analyze the demand bound of mixed-criticality industrial wireless sensor networks under the EDF policy and formulate network demand bounds in each criticality mode. Thirdly, we tighten the demand bound by analyzing carry-over jobs and classifying the number of conflicts to improve analysis accuracy. The simulations based on random network topologies demonstrate that our method can estimate network schedulability efficiently.

References

1. Chen DJ, Nixon M, Mok A (2010) Why WirelessHART. Springer, Berlin
2. Saifullab A, Gunatilaka D, Tiwari P, Sha M, Lu CY, Li B, Wu CJ, Chen YX (2015) Schedulability analysis under graph routing in WirelessHART networks. In: Real-time systems symposium (RTSS). IEEE, Piscataway, pp 165–174
3. Wu C, Gunatilake D, Saifullah A, Sha M, Tiwari PB, Lu CY, Chen YX (2016) Maximizing network lifetime of WirelessHART networks under graph routing. In: IEEE first international conference on internet-of-things design and implementation (IoTDI), pp 176–186
4. Giuseppe L (2005) Earliest deadline first. http://retis.sssup.it/~lipari/courses/rtos/lucidi/edf.pdf
5. Saifullah A, Xu Y, Lu CY, Chen YX (2010) Real-time scheduling for WirelessHART networks. In: Real-time systems symposium (RTSS). IEEE, Piscataway, pp 150–159
6. Spuri M, Reflecs P (1996) Analysis of deadline scheduled real-time systems. HAL-INRIA.
7. Baruah SK, Mok AK, Rosier LE (1990) Preemptively scheduling hard-real-time sporadic tasks on one processor. In: Real-time systems symposium (RTSS). IEEE, Piscataway, pp 182–190
8. Rox J, Ernst R (2013) Compositional performance analysis with improved analysis techniques for obtaining viable end-to-end latencies in distributed embedded systems. Int J Softw Tools Technol Transf 15(3):171–187
9. Mok AK, Feng X, Chen DJ (2001) Resource partition for real-time systems. In: Real-time technology and applications symposium (RTAS). IEEE, Piscataway, pp 75–84
10. Shin I, Lee I (2003) Periodic resource model for compositional real-time guarantees. In: Real-time systems symposium (RTSS). IEEE, Piscataway, pp 2–13
11. Ekberg P, Wang Y (2012) Bounding and shaping the demand of mixed-criticality sporadic tasks. In: The euromicro conference on real-time systems (ECRTS). IEEE, Piscataway, pp 135–144
12. Tillenius M, Larsson E, Badia RM (2015) Resource-aware task scheduling. Trans Embed Comput Syst 14(5):1–25
13. Saifullah A, Xu Y, Lu CY, Chen YX (2015) End-to-end communication delay analysis in industrial wireless networks. Trans Comput 64(5):1361–1374

14. Camilo T, Silva JS, Rodrigues A, Boavida F (2007) Gensen: a topology generator for real wireless sensor networks deployment. In: The international conference on software technologies for embedded and ubiquitous systems. Springer, Berlin, pp 436–445
15. Bini E, Buttazzo CC (2005) Measuring the performance of schedulability tests. Real-Time Syst 30(1):129–154

Chapter 4
Mixed-Criticality Scheduling for TDMA Networks

Abstract To improve the schedulability of mixed-criticality industrial wireless networks, targeted algorithms should be developed. Therefore, in this chapter, we propose a mixed-criticality scheduling algorithm. The algorithm supports centralized optimization and adaptive adjustment. It can improve both the schedulability and flexibility. We conduct extensive simulations, and the results demonstrate that the proposed scheduling algorithm significantly outperforms existing ones.

4.1 Background

Since time division multiple access (TDMA) scheduling has the high predictability, it is widely used in industrial networks [1–3]. In mixed-criticality industrial wireless networks, when there are not enough resources for all data packets, the low-criticality data packets have to be discarded. Hence, almost all of mixed criticality systems must support discarding strategies [4–7]. Previous works on single-criticality industrial wireless networks apply centralized TDMA methods to guarantee the real time performance and reliability of industrial wireless networks, e.g. [8–13]. However, the centralized TDMA methods are inflexible and difficult to cope with discarding.

Intuitively, two types of methods can be used to schedule data flows in mixed-criticality wireless networks. The first type is to schedule flows based on criticality monotonic priorities. The criticality monotonic scheduling assigns the higher priority to the important flows and schedules them first. However, this method considers the criticality as the temporality. Actually, they are not equivalent. Thus, the criticality monotonic scheduling algorithm is not suitable for mixed criticality systems. This has also been demonstrated in [14]. The second type is to use the algorithms that have been proposed for previous mixed-criticality systems, such as uniprocessor/multiprocessor systems [15–17] and networks [18–20], to solve our problem. However, industrial wireless networks are different from the previous systems. To guarantee the strict requirements on the real-time performance and reliability, the main problem to be solved is how to avoid the collision and interference between parallel data flows. Mixed-criticality uniprocessor/mulitprocessor

© The Author(s) 2023
X. Jin et al., *Mixed-criticality Industrial Wireless Networks*, Wireless Networks,
https://doi.org/10.1007/978-981-19-8922-3_4

systems only consider independent processors and do not have the interference
between parallel tasks. Mixed criticality wired networks and IEEE 802.11-based
wireless networks are based on CSMA (Carrier Sense Multiple Access) protocols,
which are unacceptable by industrial wireless networks due to the unpredictability
(We give more clarifications on the differences between our system and others
in Sect. 4.2). Therefore, previous algorithms cannot be used without modification
in mixed criticality industrial wireless networks. In this chapter, we present a
holistic scheduling solution to guarantee the real time and reliability requirements
of data flows in resource-constrained industrial wireless networks. Although some
flexible and scalable MAC protocols [21, 22] are adopted to improve the real-time
performance and reliability of networks, they are based on only local information
and cannot optimize the whole network. Therefore, our scheduling method is
implemented in the application layer. According to the generated schedules, each
network node transmits or receives packets in the MAC (Medium Access Control)
layer. The scheduling method of the application layer can manage all data flows
based on global information. Thus, it can get the optimized solution.

This chapter includes the following:

First, we propose a scheduling algorithm for mixed-criticality networks. The
scheduling algorithm not only implements the optimized global management for
all flows, but also reserves network resources for dynamic adjustments to enhance
the real time performance and reliability of important flows. It makes a trade-off
between the scheduling performance and the flexibility. Performance evaluations
demonstrate that the proposed scheduling algorithm outperforms existing ones.

Second, we present a schedulability analysis for the proposed scheduling algorithm. We analyze end-to-end delay for flows, and determine whether they are
all schedulable. Simulation results show that our schedulability analysis is more
effective than existing ones.

4.2   System Model

Industrial wireless networks must support the strict requirements on real time
performance and reliability. Therefore, we consider an industrial wireless network
as follows. It consists of a gateway and some devices. We use the node set $N =
\{n_1, n_2, \ldots\}$ to denote these nodes. The physical layer of our industrial wireless
networks is specified by the IEEE 802.15.4 protocol. It supports 16 non-overlapping
channels. However, due to external interference, not all of them can be accessed all
of the time. We denote the number of available channels as M ($1 \leq M \leq 16$). Our
network serves the flow set $F = \{f_1, f_2, \ldots\}$. Each element f_i is characterized by
$< T_i, \Pi_i, \chi_i >$. Each flow f_i periodically generates a packet at its period T_i, and
then sends it to the destination via the routing path Π_i. The relative deadline of each
packet is equal to the period T_i, i.e., a packet is released at the time t, and it must
be delivered to its destination before the time $(t + T_i + 1)$. In industrial wireless
protocols, e.g. [23, 24], periods conform to the expression

$$b \times 2^a, \tag{4.1}$$

where a is an integer value and b is the unit-period.

To keep consistent with related works on mixed criticality systems, our network also supports two criticality levels, L-crit (Low criticality) and H-crit (High criticality). The dual-criticality model can be easily extended to multi-criticality model. If the flow f_i is important, its criticality level χ_i is denoted as H. Otherwise, its criticality level χ_i is L. When the system is running in the normal mode without any exception, all flows are delivered to their destinations within deadlines. If important equipment has an exception, the corresponding data must be submitted frequently and via two paths to avoid faults on a single path. Thus, in our system model, the H-crit flows have two parameter sets: the L-crit parameters $< T_i(L), \Pi_i(L) >$ in the normal mode; the H-crit parameters $< T_i(H), \Pi_i(H) >$ in the exception mode, and $T_i(H) \leq T_i(L)$. $\Pi_i(L)$ is a path that is used by the H-crit flow in the normal mode. $\Pi_i(H)$ contains two paths that are used by the H-crit flow in the exception mode, and the two paths transmit the same packet to improve the reliability. In order to clearly distinguish these paths, they are denoted as $\Pi_i(L) = \{\pi_i^*\}$ and $\Pi_i(H) = \{\pi_i', \pi_i''\}$. The path π_i^* (and π_i', π_i'') is the set of links from the source to the destination. In this chapter, we do not consider how to select routing paths. We assume all paths have been given before generating schedules. The dynamism this chapter addresses refers to using different parameters in different modes. Transmitting a packet through the j-th link of the path π_i^* (or π_i', π_i'') is called as the transmission τ_{ij}^* (and τ_{ij}', τ_{ij}''). Each transmission has two attributes $< n_\alpha, n_\beta >$, which denote the transmission's source and destination respectively. As the constrained resources must provide enough services to H-crit flows, the L-crit flows cannot be transmitted when exceptions happen. Therefore, L-crit flows only have a parameter set $< T_i(L), \Pi_i(L) >$.

To improve the reliability of industrial networks, we adopt the TDMA scheme in the MAC layer. The network manager, which is connected to the gateway, assigns a time slot and a channel offset to each transmission. A transmission only is scheduled at the given time slot and on the given channel offset. Packets are generated periodically, and the schedules of corresponding transmissions have the same period. The schedules with the same period are organized within a *superframe* [24]. Transmitting a packet from the source to the destination has to be done in a superframe. Thus, superframes repeat themselves periodically, and then flows can be transmitted successfully. Figure 4.1a shows a simple network, which contains two flows f_1 and f_2. When the system is in normal mode, the flows use their L-crit parameters. Their periods are 8 time slots and 4 time slots, and their paths are $\{e_{52}, e_{21}\}$ and $\{e_{98}, e_{87}, e_{74}, e_{41}\}$, where e_{ij} denotes the link from the node n_i to the node n_j. Figure 4.1b shows their superframes with different periods. CH and TS denote Channel Offset and Time slot.

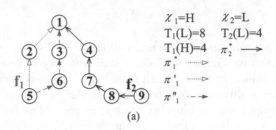

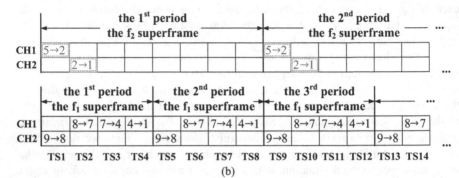

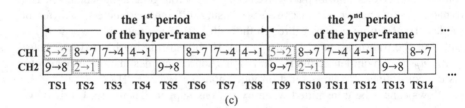

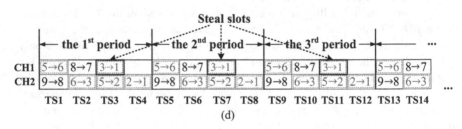

Fig. 4.1 Graph routing and superframe. (**a**) A network. (**b**) Superframes with different periods. (**c**) A hyper-frame. (**d**) The flow f_2 steal slots from the flow f_1

Two types of improper schedules will lead to transmission interference, which seriously affects the network reliability. The first type, called *node interference*, is that more than one transmissions uses the same node at the same time slot. Each node is only equipped with one transmitter. Therefore, one node cannot serve more than one transmissions at the same time. The second type is called *scheduling interference* which means that more than one transmissions is scheduled at the same time slot and on the same channel. These overlapping transmissions cannot be separated. To avoid transmission interference between different superframes, we consider all superframes as a hyper-frame whose period is the lowest common multiple of all superframes. According to the period's Expression (4.1), the hyper-period $\mathcal{T} = LCM(T_1, T_2, \ldots) = \max_{\forall f_i \in F} \{T_i\}$. Figure 4.1c shows the hyper-frame of the simple example. We only consider how to schedule flows in the first hyper-period, since after that, all schedules are repeated periodically. The network manager generates all schedules under two situations: Situation 1: when the network is deployed; and Situation 2: when the deployment is changed. Due to the requirement of industrial applications being fixed, the deployment is not often changed. Thus, the schedules may be generated several times, but not frequently. According to this schedule information, it obtains the working modes of each node at every time slot, and then delivers them to the corresponding nodes. For the schedules in Fig. 4.1c, from $TS1$ to $TS4$, working modes of the node n_2 are {receive, send, idle, idle}.

When a node intends to send a transmission of L-crit flows, it waits for a constant time and then listens to whether its channel is used. If the channel is used by H-crit flows, the node discards its transmission. Otherwise, the node sends the transmission. Note that although the node uses the carrier sense technique to determine whether an L-crit transmission is discarded or not, it is different from the CSMA scheme. For L-crit flows, the node performs carrier sense within time slots of the TDMA frame. If the L-crit transmission is not discarded, it is also scheduled based on the TDMA scheme. When a node intends to send a transmission of H-crit flows, it immediately sends it at the beginning of the assigned time slot. The scheduling algorithm assigns the proper time slot and channel for each transmission and prevents H-crit transmissions from interfering with other H-crit transmissions. Therefore, H-crit transmissions are sent directly without checking the channel. In this way, the H-crit flow can steal slots from L-crit flows when it needs more resources to cope with exceptions [25]. Note that the H-crit flow using H-crit parameters is not permitted to steal slots that are used by any other H-crit flows even if these H-crit flows are using L-crit parameters. Figure 4.1d shows an example of mixed-criticality schedules. The period of the H-crit flow f_1 is changed from 8 to 4, and the new path $\{e_{56}, e_{63}, e_{31}\}$ begins to be used. In this case, there are not enough time slots. The H-cirt transmission $3 \to 1$ (the solid line in Fig. 4.1d) steals the resource of the L-cirt transmission $7 \to 4$. Based on the stealing strategy, the dynamic adjustment can be supported.

The *schedulable* flow set is defined as follows. When the system is in the normal mode, the flow set is schedulable if all flows characterized by L-crit parameters can hit their deadlines. When there are exceptions in the system, the flow set is schedulable if all H-crit flows can hit their deadlines no matter which parameters they are using.

4.3 Problem Statement

Based on the above system model, we describe the mixed criticality scheduling problem as follows. Given the network and the flow set F, our objective is to schedule transmissions in the time slot and channel dimensions such that the flow set is schedulable.

To explain the problem more clearly, we formulate the problem as a Satisfiability Modulo Theories (SMT) specification. The transmission τ_{ij}^* (and τ_{ij}', τ_{ij}'') is assigned the s_{ij}^*-th (and s_{ij}'-th, s_{ij}''-th) time slot and the r_{ij}^*-th (and r_{ij}'-th, r_{ij}''-th) channel offset. Note that a transmission is scheduled periodically. Therefore, the transmission uses all of the time slots $s_{ij} + g \cdot T_i$ ($\forall g \in [0, \frac{T}{T_i})$) in a hyper-frame. These assignments must respect the following constraints.

(a) **Channel Offset Constraint**.

$$\forall f_i, \forall j \in [1, |\pi_i^*|], 1 \le r_{ij}^* \le M$$

For each transmission, its assigned channel offset must be in M available channels. This expression is for transmissions in the path π_i^*. Other transmissions τ_{ij}' and τ_{ij}'' in paths π_i' and π_i'' have the same constraint, and we omit them for simplicity.

(b) **Releasing Sequence Constraint**.

$$\forall f_i, \forall j \in [1, |\pi_i^*| - 1], s_{i,j}^* < s_{i,j+1}^*$$

In a routing path, the transmission $\tau_{i,j+1}$ is released after the transmission $\tau_{i,j}$ is scheduled. We still omit paths π_i' and π_i''.

(c) **Real Time Constraint**.

$$\forall f_i, 1 \le s_{i,|\pi_i^*|}^* \le T_i(L)$$

All transmissions cannot miss deadlines. Likewise, $s_{i,|\pi_i'|}'$ and $s_{i,|\pi_i''|}''$ have the same constraint.

(d) **Interference Constraint**. Assigning resources to transmissions must prevent the happening of node interference and scheduling interference. We use $\delta(\tau_a, \tau_b)$ to denote whether there exists interference between τ_a and τ_b,

$$\delta(\tau_a, \tau_b) = (\tau_a \cap \tau_b = \emptyset)?(\eta(s_a, s_b) \wedge (r_a = r_b)) : \eta(s_a, s_b),$$

where $\eta(s_a, s_b) = \bigvee\limits_{\forall h \in [0, \frac{T}{T_a}), \forall k \in [0, \frac{T}{T_b})} (s_a + h \cdot T_a = s_b + k \cdot T_b)$ means whether the assigned time slots of τ_a and τ_b overlap each other. If the two transmissions do not use the same node, i.e., $\tau_a \cap \tau_b = \emptyset$, then they can be scheduled at different time slots or on the different channel offsets. Otherwise, there exists node interference and they cannot be scheduled at the same time slot. The transmissions of the H-crit flow f_i are classified into three sets $\Gamma_i^* = \{\tau_{ij}^* | \forall j \in [1, |\pi_i^*|]\}$, $\Gamma_i' = \{\tau_{ij}' | \forall j \in [1, |\pi_i'|]\}$ and $\Gamma_i'' = \{\tau_{ij}'' | \forall j \in [1, |\pi_i''|]\}$. For the L-cirt flow f_i, $\Gamma_i' = \Gamma_i'' = \emptyset$, and then $\forall f_i \in F, \Gamma_i = \Gamma_i^* \cup \Gamma_i' \cup \Gamma_i''$. Thus, the interference constraints in the normal mode and exception mode are as follows.

(d.1) Normal mode

$$\forall \tau_a, \tau_b \in \bigvee_{\forall f_i \in F} \Gamma_i^*, \delta(\tau_a, \tau_b) = 0$$

(d.2) Exception mode

$$\forall f_i, f_g \in F, \chi_i = \chi_g = H, \forall \tau_a \in \Gamma_i, \forall \tau_b \in \Gamma_g, \delta(\tau_a, \tau_b) = 0$$

The mixed criticality scheduling problem is NP-hard [26]. Our SMT specification can be solved by some solvers, such as Z3 [27] and Yices [28]. These solvers can find satisfying assignments for quite many problems, and their solutions have been an excellent standard to evaluate the effectiveness of other methods [29]. However, the running time may be unacceptable for complex networks and flow sets. Therefore, we propose a heuristic scheduling algorithm in Sect. 4.4 to solve the problem.

4.4 Scheduling Algorithm

In this section, we first introduce how to schedule transmissions, and then, based on these schedules, we determine working modes of each node at every time slot.

4.4.1 A Slot-Stealing Scheduling Algorithm

We propose a slot-stealing scheduling algorithm based on RM (StealRM). The proposed StealRM optimizes the solution according to the global information, and permits transmissions to share the same resource when the transmissions have different levels of criticality. Hence, the schedules can be adaptively adjusted based on the requirements of H-crit flows.

The proposed StealRM is shown in Algorithm 4.1. Each flow is assigned as the RM priority. If two flows have the same RM priority, the flow with the smaller ID has the higher priority. The transmission's priority is equal to its flow's priority. The set R contains all of schedulable transmissions (lines 1 and 19), and the set R' denotes released transmissions at the current time slot (line 3). At every time slot t, we first sort elements of R' according to the decreasing order of priorities, and τ_1

in the set R' has the highest priority (line 4). Then, for each transmission τ_a in the set R', we check whether it can be scheduled at the current time slot without any interference (lines 7–23). Let $\mathcal{F}(\tau_a)$ denote the flow that the transmission τ_a belongs to (line 6). The set Y_t^{HL} contains the transmissions that have been scheduled at the time slot t and belong to H-crit flows with L-crit parameters. The sets Y_t^H and Y_t^L correspond to those in H-crit flows with H-crit parameters and L-crit flows, respectively. The transmissions in the set Y' and the transmission τ_a cannot steal slots from each other. According to the criticality level of τ_a, the set Y' is assigned different transmissions (lines 7–13). If the transmission τ_a belongs to an H-crit flow with H-crit parameters, then it cannot steal slots from other H-crit transmissions (lines 7–8). Y^H and Y^{HL} may contain the transmissions belonging to the same flow with τ_a. These transmissions do not interfere the scheduling of τ_a. Thus, the set $\{\forall \tau_{ig}^*\}$ needs to be excluded from Y^H and Y^{HL} (line 8). Similarly, if the transmission τ_a belongs to an H-crit flow with L-crit parameters, then it cannot steal slots from any other transmissions (lines 9 and 10). If the transmission τ_a belongs to an L-crit flow, then its slots cannot be stolen by L-crit flows and H-crit flows with L-crit parameters (lines 11 and 13). When there is no node interference between τ_a and Y', and at least one channel is idle (line 14), the transmission τ_a can be scheduled at this current time slot. $\Theta(Y')$ denotes the channels that have been used by Y'. However, if the current time slot has exceeded its deadline, the flow set is unschedulable (lines 15 and 16). Otherwise, the time slot and channel offset of the transmission τ_a are assigned (line 18), and the schedulable transmission set R and the scheduled transmission set Y_t^H (Y_t^L and Y_t^{HL}) are updated (lines 19–26).

The number of iterations of the **for** loop in line 2 and the **for** loop in line 5 is $O(|\mathcal{T}|)$ and $O(|\Gamma|)$, respectively. The complexity of line 4, line 14 and line 21 is $O(|\Gamma|log|\Gamma|)$, $O(|\Gamma|)$ and $O(\frac{\mathcal{T}}{T_{min}})$, respectively. Therefore, the time complexity of Algorithm 4.1 is $O(|\mathcal{T}||\Gamma|^2 \frac{\mathcal{T}}{T_{min}})$.

4.4.2   Node Working Mode

Nodes have three working modes, including transmit mode ($\mathbb{S}$), receive mode ($\mathbb{R}$) and idle mode. We use $w_{\alpha,t}^H = <\mathbb{S}\,(\text{or }\mathbb{R}), r_a>$ to denote that at the time slot t the node n_α transmits (or receives) H-crit flows on the channel r_a. Similarly, $w_{\alpha,t}^L$ denotes that the node n_α serves L-crit flows. Algorithm 4.2 determines the working mode for each node. For each transmission, we have assigned a time slot and a channel offset in Algorithm 4.1. According to the assignments, the working modes of the sender node and receiver node of the transmission can be obtained (lines between 4 and 10). The time complexity of Algorithm 4.2 is $O(|\Gamma|\frac{\mathcal{T}}{T_{min}})$.

Note that a node may serve two flows at the same time slot, but the two flows must have different criticality levels. Otherwise, node interference occurs. At the beginning of the time slot t, the node works in mode $w_{\alpha,t}^H$. Then, in a constant time if it needs to send an H-crit flow or has received a flow, it continues working in the

---

Algorithm 4.1 StealRM

Input: the flow set F
Output: the scheduling results $\forall s_a$ and $\forall r_a$;
1: the schedulable transmission set $R \leftarrow \{\tau_{i1}^*, \tau_{i1}', \tau_{i1}'' | \forall f_i \in F\}$;
2: **for** $\forall t \in [1, \mathcal{T}]$ **do**
3:     $R' \leftarrow R$;
4:     sort R' according to the decreasing order of priorities;
5:     **for** each a from 1 to $|R'|$ **do**
6:         $i \leftarrow \mathcal{F}(\tau_a)$;
7:         **if** $\chi_i == H$ and $\tau_a \in \Gamma_i' \cup \Gamma_i''$ **then**
8:             $Y' \leftarrow \bigcup\limits_{\forall h \in [0, \frac{\mathcal{T}}{T_i(H)})} (Y_{t+T_i(H) \times h}^H \cup Y_{t+T_i(H) \times h}^{HL}) - \{\forall \tau_{ig}^*\}$;
9:         **else if** $\chi_i == H$ and $\tau_a \in \Gamma_i^*$ **then**
10:           $Y' \leftarrow (\bigcup\limits_{\forall h \in [0, \frac{\mathcal{T}}{T_i(L)})} Y_{t+T_i(L) \times h}^L) \cup (\bigcup\limits_{\forall h \in [0, \frac{\mathcal{T}}{T_i(H)})} (Y_{t+T_i(H) \times h}^H \cup Y_{t+T_i(H) \times h}^{HL})) - \{\forall \tau_{ig}', \tau_{ig}''\}$;
11:         **else**
12:           $Y' \leftarrow (\bigcup\limits_{\forall h \in [0, \frac{\mathcal{T}}{T_i(L)})} Y_{t+T_i(L) \times h}^L) \cup (\bigcup\limits_{\forall h \in [0, \frac{\mathcal{T}}{T_i(H)})} Y_{t+T_i(H) \times h}^{HL})$;
13:         **end if**
14:         **if** $\bigwedge\limits_{\forall \tau_b \in Y'} (\tau_a \cap \tau_b \neq \emptyset)$ and $|\Theta(Y')| < M$ **then**
15:             **if** t exceeds the deadline of f_i **then**
16:                 **return** unschedulable;
17:             **end if**
18:             $s_a \leftarrow t; r_a \leftarrow$ a random channel that is not in $\Theta(Y')$;
19:             $R \leftarrow R - \{\tau_a\}+$ the next transmission of τ_a;
20:             **if** $\chi_i == H$ and $\tau_a \in \Gamma_i' \cup \Gamma_i''$ **then**
21:                 $\forall h \in [0, \frac{\mathcal{T}}{T_i(H)}), Y_{t+T_i(H) \times h}^H \leftarrow Y_{t+T_i(H) \times h}^H + \{\tau_a\}$;
22:             **else if** $\chi_i == H$ and $\tau_a \in \Gamma_i^*$ **then**
23:                 $\forall h \in [0, \frac{\mathcal{T}}{T_i(L)}), Y_{t+T_i(L) \times h}^{HL} \leftarrow Y_{t+T_i(L) \times h}^{HL} + \{\tau_a\}$;
24:             **else**
25:                 $\forall h \in [0, \frac{\mathcal{T}}{T_i(L)}), Y_{t+T_i(L) \times h}^L \leftarrow Y_{t+T_i(L) \times h}^L + \{\tau_a\}$;
26:             **end if**
27:         **end if**
28:     **end for**
29: **end for**
30: **return** $\forall s_a$ and $\forall r_a$;

---

same mode at this time slot. Otherwise, it works in mode $w_{\alpha,t}^L$. However, when its mode $w_{\alpha,t}^L$ is $\mathbb{S}$, it must determine whether the assigned channel is clear or not before it sends the flow. If the channel has been occupied by H-crit flows, the flow has to be discarded. The switch time between different modes is very short compared with a time slot. For example, the switch time of the transceiver CC2420 is just $200\,\mu s$ while a time slot is 10 ms. Generally, at a time slot, most nodes only serve one flow or are idle, while only a few nodes serve two flows.

Algorithm 4.2 Working mode

Input: the scheduling results $\forall s_a$ and $\forall r_a$
Output: all $w_{*,*}^L$ and $w_{*,*}^H$
1: all $w_{*,*}^L$ and $w_{*,*}^H$ are initiated as idle mode;
2: **for** $\forall \tau_a \in \Gamma$ **do**
3:      $i \leftarrow \mathcal{F}(\tau_a)$; $< n_\alpha, n_\beta >$ are the sender and receiver of τ_a;
4:      **if** $\chi_i == H$ **then**
5:          $\forall h \in [0, \frac{\mathcal{T}}{T_i(H)})$, $w_{\alpha, s_a + T_i(H) \times h}^H \leftarrow < \mathbb{S}, r_a >$;
6:          $\forall h \in [0, \frac{\mathcal{T}}{T_i(H)})$, $w_{\beta, s_a + T_i(H) \times h}^H \leftarrow < \mathbb{R}, r_a >$;
7:      **else**
8:          $\forall h \in [0, \frac{\mathcal{T}}{T_i(L)})$, $w_{\alpha, s_a + T_i(L) \times h}^L \leftarrow < \mathbb{S}, r_a >$;
9:          $\forall h \in [0, \frac{\mathcal{T}}{T_i(L)})$, $w_{\beta, s_a + T_i(L) \times h}^L \leftarrow < \mathbb{R}, r_a >$;
10:     **end if**
11: **end for**
12: **return** all $w_{*,*}^L$ and $w_{*,*}^H$;

4.5   Scheduling Analysis

In this section, we analyze the worst case end-to-end delay for each flow and use the delay to test the schedulability of the flow set. If the worst case delay of all flows does not exceed deadlines, the flow set is schedulable. For the sake of simplicity, we first explain how to compute the worst case delay in single-criticality networks (in Sect. 4.5.1) and then extend it to mixed-criticality networks (in Sect. 4.5.2).

4.5.1   Analyzing Method for Single-Criticality Networks

Besides transmitting time, the end-to-end delay is introduced by the interference from higher priority flows. Therefore, in Sect. 4.5.1.1, we present the analyzing method of the total interference. In Sect. 4.5.1.2 we distinguish the different types of interference and compute the worst case delay.

4.5.1.1   Total Interference

During the time interval between the release and completion of the flow f_k, all the active transmissions that belong to the higher priority flows may have node interference or scheduling interference to the flow f_k. Therefore, in the worst case, the total interference is equal to the number of those higher-priority transmissions. The method of computing the workload in a period has been proposed in multi-processor systems [30]. The mapping between the multiprocessor system model and the network model has been explained in the work [31], in which a channel corresponds to a processor and a flow is scheduled as a task. Therefore, we propose

our analyzing method based on the work [30], which is the start-of-the-art analysis
for multiprocessor systems. To make this chapter self-contained, we first simply
introduce the method of multiprocessor systems, and then present our method.

For the simplicity of expression, the multiprocessor system uses the same
notations as our network model. For multiprocessor systems, the calculation of the
worst case delay of the task f_k is based on the *level-k busy period* (as shown in
Definition 4.1).

Definition 4.1 (Level-k Busy Period for Multiprocessor Systems) The level-k
busy period is the time interval $[t_0, t_k)$, in which t_k is the finish time of the task
f_k, and t_0 satisfies the following conditions:

1. $t_0 < t_r$ where t_r is the release time of the task f_k.
2. $\forall t \in [t_0, t_r]$, at the time t, all processors are occupied by higher-priority tasks.
3. $\forall t < t_0, \exists t' \in [t, t_0]$, at the time t', at least one processor is occupied by lower-
   priority tasks.

If there is no t_0 that satisfies all conditions, then $t_0 = t_r$.

The level-k busy period is determined by the workload of all higher-priority tasks.
The set $\bar{P}(f_k)$ contains the tasks with higher priority than the task f_k. If the task
f_i ($f_i \in \bar{P}(f_k)$) has a job that is released earlier than the level-k busy period and
its deadline is in the busy period, then the task f_i has the carry-in workload in the
level-k busy period. Otherwise, the task has no carry-in workload. The two types of
workload are presented as follows, and the length of the level-k busy period is x.

1. In the level-k busy period, if the task f_i has no carry-in workload, the upper
   bound of its workload is

$$W_k^{NC}(f_i, x) = \left\lfloor \frac{x}{T_i} \right\rfloor \cdot c_i + \min\{x \mod T_i, c_i\},$$

   where c_i is the execution time of the task f_i.
2. If the task f_i has the carry-in workload, the upper bound of its workload is

$$W_k^{CI}(f_i, x) = \left\lfloor \frac{\max\{x - c_i, 0\}}{T_i} \right\rfloor \cdot c_i + c_i + \alpha,$$

   where $\alpha = \min\{\max\{\max\{x - c_i, 0\} - (T_i - D_i), 0\}, c_i - 1\}$ and D_i is the worst
   case delay of the task f_i.

Based on the upper bounds of workload, two types of interference of the task f_i
to the task f_k are as follows:

$$I_k^{NC}(f_i, x) = \min\{\max\{W_k^{NC}(f_i, x), 0\}, x - c_k + 1\},$$

$$I_k^{CI}(f_i, x) = \min\{\max\{W_k^{CI}(f_i, x), 0\}, x - c_k + 1\}.$$

Therefore, the total interference suffered by the task f_k is

$$\Omega_k(x, \bar{P}^{NC}(f_k), \bar{P}^{CI}(f_k)) = \sum_{\forall f_i \in \bar{P}^{NC}(f_k)} I_k^{NC}(f_i, x) + \sum_{\forall f_i \in \bar{P}^{CI}(f_k)} I_k^{CI}(f_i, x),$$

where $\bar{P}^{NC}(f_k)$ and $\bar{P}^{CI}(f_k)$ denotes the set of tasks without carry-in workload and the set of tasks with carry-in workload, respectively. In a busy period, at most $M - 1$ higher-priority tasks have carry-in workload. Therefore, the set $\bar{P}^{CI}$ contains $M - 1$ tasks that have maximal values of $I_k^{CI}(f_i, x) - I_k^{NC}(f_i, x)$. Other tasks are in the set $\bar{P}^{NC}$.

In the following, we propose our analyzing method. Industrial wireless networks apply strict periodic schedules based on superframes, which can reduce system complexity and run time overhead. While in multiprocessor systems and previous works about wireless networks, schedules are variable, i.e., the assigned time slots to a task (or a flow) are non-periodic, so our workload bounds are not the same as previous ones. Our workload bounds are computed with Theorem 4.1. Definition 4.2 defines the level-k busy period in the network.

Definition 4.2 (Level-k Busy Period for Networks) The level-k busy period is the time interval $[t_0, t_k)$, in which t_k is the finish time of the flow f_k and t_0 satisfies the following conditions:

1. $t_0 < t_r$ where t_r is the release time of the flow f_k.
2. $\forall t \in [t_0, t_r]$, at the time t, all channels are occupied by higher-priority flows or there exists node interference between the scheduled flows and the flow f_k.
3. $\forall t < t_0, \exists t' \in [t, t_0]$, at the time t', there is no node interference and at least one channel is occupied by lower-priority flows or idle.

If there is no t_0 that satisfies all conditions, then $t_0 = t_r$.

Theorem 4.1 *The workload bounds can be computed with*

$$W_k^{NC}(f_i, x) = W_k^{CI}(f_i, x) = \left\lfloor \frac{x}{T_i} \right\rfloor \cdot c_i + \min\{x \mod T_i, c_i\}, \tag{4.2}$$

where c_i is the number of hops in the path π_i, i.e. $c_i = |\pi_i|$.

Proof of Theorem 4.1 The computation of the non-carry-in workload $W_k^{NC}(f_i, x)$ is shown in Fig. 4.2a. There are $\left\lfloor \frac{x}{T_i} \right\rfloor$ complete periods and a scheduling window ($x \mod T_i$). In the scheduling window, at most c_i workloads exist. Therefore, the expression of the non-carry-in workload is shown as Eq. (4.2).

In the following, we compute W_k^{CI} as shown in Fig. 4.2b. The notations A and B denote the two incomplete periods, respectively. We know that $A < T_i$, $B < T_i$ and $A + B = (x \mod T_i)$ or $(x \mod T_i + T_i)$. We discuss the two cases as follows.

Case 1: $A + B = x \mod T_i$ We draw out the windows A and B in Fig. 4.3a. We consider four different value ranges of the windows A and B as shown in Table 4.1,

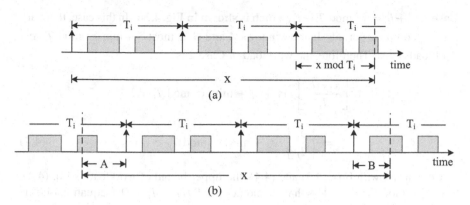

Fig. 4.2 Illustration of Theorem 4.1. (a) $W_k^{NC}(f_i, x)$. (b) $W_k^{CI}(f_i, x)$

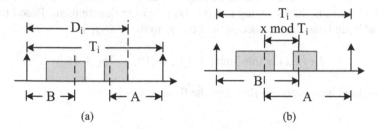

Fig. 4.3 Computation of W_k^{CI}. (a) $A + B = x \mod T_i$. (b) $A + B = x \mod T_i + T_i$

in which if $A \geq T_i - D_i$ and $B \geq D_i$, it is Case 2. If $A < T_i - D_i$, then there is no workload in A. If $B \geq D_i$, then all execution time c_i must be the available workload. In this case, the workload can also be expressed as $\min\{B, c_i\}$. Therefore, only if $A < T_i - D_i$, the workload is $\min\{B, c_i\}$. If $A \geq T_i - D_i$ and $B < D_i$, the time interval $T_i - D_i$ does not contain any workload. Therefore, the available window $A + B$ is equal to $(x \mod T_i) - (T_i - D_i)$.

In Case 1, we can get that the total workload is

$$\left\lfloor \frac{x}{T_i} \right\rfloor \cdot c_i + C1. \tag{4.3}$$

The notation $C1$ denotes the workload in the incomplete period as shown in Table 4.1. It is equal to $\min\{B, c_i\}$ or $\min\{x \mod T_i - (T_i - D_i), c_i\}$.

Table 4.1 The workload in the incomplete period under different value ranges of A and B

Workload	$A < T_i - D_i$	$A \geq T_i - D_i$
$B \geq D_i$	c_i	Case 2
$B < D_i$	$\min\{B, c_i\}$	$\min\{x \mod T_i - (T_i - D_i), c_i\}$
C1	$\min\{B, c_i\}$	$\min\{x \mod T_i - (T_i - D_i), c_i\}$

Case 2: $A+B = x \mod T_i+T_i$, **which is shown in Fig. 4.3b** In this case, there are $\left\lfloor \frac{x-T_i}{T_i} \right\rfloor$ complete periods. In the windows A and B, at most $c_i+\min\{x \mod T_i, c_i\}$ workloads exist. Therefore, the workload of Case 2 is

$$\left\lfloor \frac{x - T_i}{T_i} \right\rfloor \cdot c_i + c_i + \min\{x \mod T_i, c_i\}$$

$$\Rightarrow \left\lfloor \frac{x}{T_i} \right\rfloor \cdot c_i + \min\{x \mod T_i, c_i\}. \tag{4.4}$$

Comparing with Eqs. (4.3) and (4.4), the upper bound of workload is Eq. (4.4). Since $(x \mod T_i)$ is not less than B and $(x \mod T_i) - (T_i - D_i)$, equation (4.4) is the same as Eq. (4.2). The theorem holds.   □

Due to the two types of workload having the same computing formula, we do not distinguish them in the following and use $W_k(f_i, x)$ to denote them. Based on the workload bound, the interference of the flow f_i to the flow f_k is

$$I_k(f_i, x) = \min\{\max\{W_k(f_i, x), 0\}, x - c_k + 1\}.$$

Thus, the total interference suffered by the flow f_k is

$$\Omega_k^{total}(x, \bar{P}(f_k)) = \sum_{\forall f_i \in \bar{P}(f_k)} I_k(f_i, x).$$

4.5.1.2 Worst Case Delay in Single-Criticality Networks

$\Omega_k^n(x, \bar{P}(f_k))$ and $\Omega_k^s(x, \bar{P}(f_k))$ denote node interference and scheduling interference suffered by the flow f_k in the level-k busy period. If there exists a node interference at a time slot, the flow f_k cannot be transmitted at this time slot, no matter how many channels are idle, i.e., the flow f_k is delayed one time slot due to the node interference. However, only when M transmissions are scheduled at a time slot, does the flow f_k suffer scheduling interference and is delayed for one time slot. In the worst case, all the node interference and scheduling interference will introduce a delay to the flow f_k. Therefore, the worst case delay is

$$\Omega_k^n(x, \bar{P}(f_k)) + \left\lfloor \frac{\Omega_k^s(x, \bar{P}(f_k))}{M} \right\rfloor + c_k. \tag{4.5}$$

From Eq. (4.5), we know that node interference introduces more delay. Since the sum of node interference and scheduling interference is $\Omega_k^{total}(x, \bar{P}(f_k))$, so when as much as possible node interference occurs, the end-to-end delay is the worst case.

The upper bound of node interference introduced by h consecutive hops of the flow f_i to the flow f_k is computed as

$$R_{k,i}(h) = \max_{\forall a \in [1, c_i - h]} \{|\{\tau_{iy} | \forall \tau_{iy}, y \in [a, a+h], \exists \tau_{kz} \text{ such that } \tau_{iy} \cap \tau_{kz} \neq \emptyset\}|\}.$$

Thus, the workload introduced by transmissions that have node interference is

$$W_k^n(f_i, x) = \left\lfloor \frac{x}{T_i} \right\rfloor \cdot R_{k,i}(c_i) + R_{k,i}(\min\{x \mod T_i, c_i\}).$$

Then,

$$I_k^n(f_i, x) = \min\{\max\{W_k^n(f_i, x), 0\}, x - c_k + 1\},$$

and

$$\Omega_k^n(x, \bar{P}(f_k)) = \sum_{\forall f_i \in \bar{P}(f_k)} I_k^n(f_i, x).$$

Then, we can get that the worst case delay of the flow f_k in the single-criticality network is

$$D_k = \Omega_k^n(x, \bar{P}(f_k)) + \left\lfloor \frac{\Omega_k^{total}(x, \bar{P}(f_k)) - \Omega_k^n(x, \bar{P}(f_k))}{M} \right\rfloor + c_k.$$

From the definition of the level-k busy period, we know that the length x is the upper bound of the delay D_k (shown in Theorem 4.2).

Theorem 4.2 *For the flow f_k and the level-k busy period, the following holds:*

$$x \geq D_k.$$

Proof of Theorem 4.2 We assume by contradiction that $x < D_k$. From the definition of the level-k busy period (Definition 4.2), we know that the finish times of the busy period and the flow f_k are the same, and t_0 must be less than (the first condition) or equal to t_r (when t_0 does not satisfy at least one condition). If $x < D_k$, then $t_r < t_0$ as shown in Fig. 4.4. It is not consistent with the definition. The above assumption does not hold. $\square$

According to Theorem 4.2, the solution of Eq. (4.6) is the upper bound of end-to-end delay D_k.

$$x = \Omega_k^n(x, \bar{P}(f_k)) + \left\lfloor \frac{\Omega_k^{total}(x, \bar{P}(f_k)) - \Omega_k^n(x, \bar{P}(f_k))}{M} \right\rfloor + c_k \qquad (4.6)$$

Fig. 4.4 Illustration of
Theorem 4.2

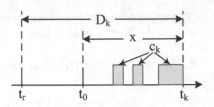

Equation (4.6) can be solved by the iterative fixed point search [32]. The iterative calculation of x starts with $x = c_k$; until the value of x does not change.

4.5.2   Mixed-Criticality Scheduling Analysis

In dual-criticality networks, there are three types of worst case delay.

1. D_k^L: the worst case end-to-end delay of the L-crit flow.
2. D_k^{HL}: the worst case end-to-end delay of the H-crit flow with the L-crit parameter.
3. D_k^H: the worst case end-to-end delay of the H-crit flow with the H-crit parameter

We use $D(x, Q, c)$ to denote $\Omega_k^n(x, Q) + \left\lfloor \frac{\Omega_k^{total}(x, Q) - \Omega_k^n(x, Q)}{M} \right\rfloor + c$. The methods of computing these types of delays are similar. The only difference is that higher-priority flows they suffered are different, i.e., their interference sets Q are different. H-crit flows have multiple paths. These paths suffer different interference and cause different delays. Therefore, we use sub-flows f_k^*, f_k' and f_k'' to distinguish them.

If there are H-crit flows with H-crit parameters in networks, L-crit flows can be discarded. Therefore, when we compute the delay D_k^L, all flows have L-crit parameters. Thus, $D_k^L = D(x, Q_k^L, c_k^*)$, where $Q_k^L = \{f_i^* | \forall f_i^*, T_i(L) < T_k(L)\}$ and $c_k^* = |\pi_k^*|$.

Similarly, for H-crit flows with L-crit parameters, the interference set is $Q_k^{HL} = \{f_i', f_i'' | \forall f_i', \forall f_i'', \chi_i = H, T_i(H) < T_k(L)\} \cup \{f_i^* | \forall f_i^*, T_i(L) < T_k(L)\}$. Thus, $D_k^{HL} = D(x, Q_k^{HL}, c_k^*)$, where $c_k^* = |\pi_k^*|$.

An H-crit flow with its H-crit parameter suffers the interference from H-crit flows with H-crit parameters. The H-crit flow has two sub-flows f_k' and f_k''. For these sub-flows, their interference set is $Q_k^H = \{f_i', f_i'' | \forall f_i', \forall f_i'', \chi_i = H, T_i(H) < T_k(H)\} \cup \{f_i^* | \forall f_i^*, \chi_i = H, T_i(L) < T_k(H)\}$ and $c_k' = |\pi_k'|$, $c_k'' = |\pi_k''|$. Thus, $D_k'^H = D(x, Q_k^H, c_k')$ and $D_k''^H = D(x, Q_k^H, c_k'')$, and then $D_k^H = \max\{D_k'^H, D_k''^H\}$.

According to the above delays, the schedulability test is as follows. For the L-crit flow f_k, if $D_k^L \leq T_k(L)$, it is schedulable; otherwise, unschedulable. For the H-crit flow f_k, if $D_k^{HL} \leq T_k(L)$ and $D_k^H \leq T_k(H)$, it is schedulable; otherwise, unschedulable. If all flows in a flow set are schedulable, the set is schedulable.

4.6 Performance Evaluations

In this section, we conduct experiments to evaluate the performance of our proposed methods.

4.6.1 Scheduling Algorithm

We consider three comparison methods: (1) **SMT** uses the Z3 solver [27], which is a high-performance solver being developed at Microsoft Research and whose solution has been regarded as an excellent standard, to solve our SMT specification (Sect. 4.3); (2) **noStealRM** applies the RM priority and does not allow slots to be stolen; (3) **StealCM** allows slots to be stolen and applies the criticality monotonic priority. Our method is **StealRM**. The performance metric we used is the *Schedulable ratio*, which is defined as the percentage of flow sets for which a scheduling algorithm can find a schedulable solution.

We randomly generate a number of test cases to evaluate these methods. For each test case, the number of channels M and the number of nodes $|N|$ are given. According to the suggestion in the work [33], these nodes are placed randomly in the square area A, and $A = \frac{|N|d^2\sqrt{27}}{2\pi}$, where the transmitting range d is 40 meters. Except for the gateway, each node has a data flow from itself to the gateway or vice versa. There are two necessary schedulability conditions for flow sets: (1) the network utilization U is not larger than 1; (2) the utilization of each node is not larger than 1. If a flow set does not satisfy the two conditions, it cannot be scheduled. Thus, in order to make flow sets available, we specify the network utilization $U(U < 1)$, and use the method UUniFast [34] to assign the utilization u_i for each flow, where $U = \sum_{\forall f_i \in F} u_i$. Then, if the flow set can satisfy condition (2), it is an available flow set. Otherwise, discard it, and repeat the process until an available set is found. The period of each flow can be obtained according to $T_i = \frac{c_i}{u_i}$. The high-crit probability of the flows is controlled by the parameter ρ. Routing paths are selected randomly.

In order to make test cases solvable by the Z3 solver, the parameters are set as $\rho = 0.3$, $M = 2$ and $U = 0.8$. For each configuration, 100 test cases are checked using the four algorithms. Figure 4.5 shows their schedulable ratios. Our algorithm StealRM is close to the result of Z3. In these simple test cases, the method StealCM has similar results with our algorithm StealRM. Figure 4.6 shows the average execution time of solvable test cases in Fig. 4.5. When the number of nodes is 25, the execution time of the method SMT is about 16.5 minutes. We also use the method SMT to solve the network with 30 nodes, but cannot get the result within 3 hours. Except for the method SMT, the execution time of other methods is not more than 10 milliseconds. Therefore, from the perspective of execution time, heuristic algorithms are significantly more efficient than the method SMT.

Fig. 4.5 Schedulability
comparison among all
methods

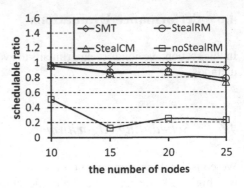

Fig. 4.6 Execution time
comparison among all
methods

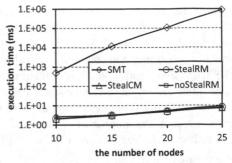

Since the execution time of the method SMT is too long, the following experiments do not contain it. Figure 4.7 shows the schedulable ratios of the three scheduling algorithms. For each point in the figure, 500 test cases are randomly generated. From the figure, we can know that our algorithm StealRM has the highest schedulable ratio no matter with which parameters, while the algorithm noStealRM has the worst result. Therefore, the stealing mechanism can significantly improve the algorithm's performance. Our algorithm StealRM has better performance than the algorithm StealCM, especially when the node numbers are higher. This demonstrates that: (1) the priority should correspond to the urgency, but not the importance, while the stealing mechanism reflects the importance; (2) the urgency and the importance have to be distinguished, except in very small networks. Comparing these subfigures, we observe that schedulable ratios decrease with the increases of ρ, $|N|$, U and M. The reasons are as follows. An H-crit flow can be regarded as two L-crit flows. Thus, a larger value of the parameter ρ leads to more flows, which are hard to schedule. A test case contains $|N| - 1$ flows. Likewise, the larger $|N|$ makes scheduling hard. The network utilization U corresponds to the network workload. Heavy workloads lead to scheduling failures. Note that compared with Fig. 4.7a, d has three additional channels, but its schedulable ratios decrease. Because the two subfigures generate test cases according to the respective numbers of channels. Their test cases are different. Although the number of channels increases, the utilization is not changed. When the utilization U is constant, with the increase of the number of channels M, the packets that need to be transmitted increase. The increased packets

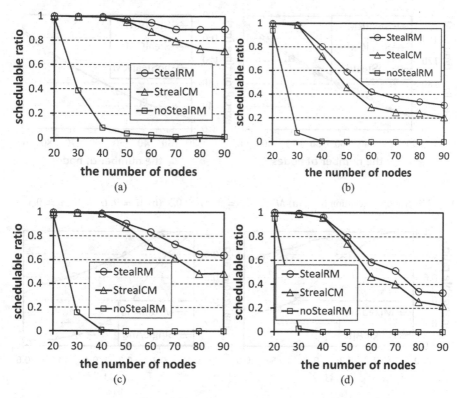

Fig. 4.7 Schedulability comparison among StealRM, StealCM and noStealRM. (**a**) $M = 6, U = 0.5, \rho = 0.3$. (**b**) $M = 6, U = 0.5, \rho = 0.4$. (**c**) $M = 6, U = 0.6, \rho = 0.3$. (**d**) $M = 9, U = 0.5, \rho = 0.3$

will introduce more interference, which has a negative impact on the scheduling performance. Therefore, Fig. 4.7d has a lower schedulable ratio than Fig. 4.7a.

Figure 4.8 shows the average execution time of Fig. 4.7. As the results are similar, we only show two subfigures for Fig. 4.7a, d. Compared with our algorithm StealRM, the algorithms StealCM and noStealRM need more time to find feasible solutions. Therefore, their execution time slightly increases. From the figure, we know that our algorithm StealRM does not introduce extra time costs. For the three algorithms, the execution time increases with the increase of the number of nodes, since more data flows need to be scheduled.

4.6.2 Analyzing Method

The comparison method is **SingleAna**, in which flow sets are tested using the single-criticality analysis. Our mixed-criticality analysis method is **MixedAna**. The performance metrics are the *analyzable ratio* (the percentage of flow sets which

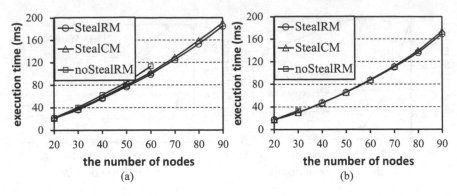

Fig. 4.8 Average execution time. (**a**) $M = 6, U = 0.5, \rho = 0.3$. (**b**) $M = 9, U = 0.5, \rho = 0.3$

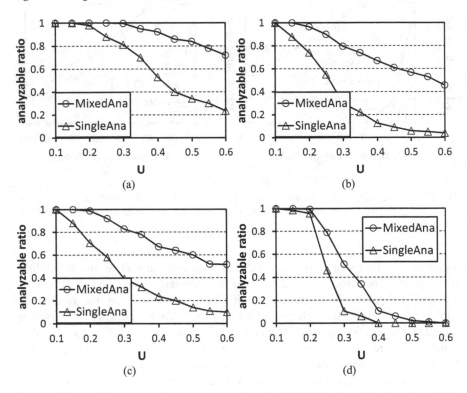

Fig. 4.9 Schedulability comparison among analyzing algorithms. (**a**) $|N| = 20, M = 6, \rho = 0.1$. (**b**) $|N| = 20, M = 6, \rho = 0.3$. (**c**) $|N| = 20, M = 9, \rho = 0.1$. (**d**) $|N| = 60, M = 6, \rho = 0.1$

are tested as schedulable by an analyzing method) and the *pessimism ratio* (the proportion of analyzed delay to the delay observed in StealRM). Figure 4.9 shows the comparison of analyzable ratios. For each point, 500 test cases are analyzed. Our method MixedAna outperforms SingleAna. The analyzable ratios decrease with

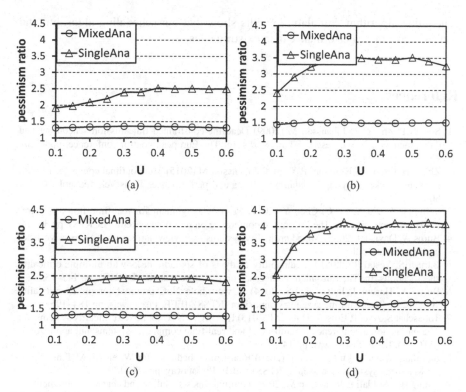

Fig. 4.10 Delay comparison with StealRM being used as the baseline. (**a**) $|N| = 20$, $M = 6$, $\rho = 0.1$. (**b**) $|N| = 20$, $M = 6$, $\rho = 0.3$. (**c**) $|N| = 20$, $M = 9$, $\rho = 0.1$. (**d**) $|N| = 60$, $M = 6$, $\rho = 0.1$

the increase of these parameters. The reasons are similar to those in Fig. 4.7. The increases of U and M lead to more packets, and the increases of $|N|$ and ρ lead to more flows. These will cause more interference. Thus, the analysis introduces more pessimism, and the analyzable ratios decrease. Figure 4.10 shows the pessimism ratios of experiments in Fig. 4.9. The pessimism ratios of MixedAna are less than 2, while the pessimism ratios of SingleAna are all larger than 2. This is because the interference that does not exist between H-crit and L-crit flows is eliminated in MixedAna.

4.7 Summary

Mixed-criticality data flows coexist in advanced industrial applications. They share the network resource, but their requirements for the real time performance and reliability are different. In this chapter, we propose a scheduling algorithm to guarantee their different requirements, and then analyze the schedulability of this scheduling algorithm. Simulation results show that our scheduling algorithm and analysis have more performance than existing ones.

References

1. Soldati P, Zhang H, Johansson M (2009) Deadline-constrained transmission scheduling and data evacuation in WirelessHART networks. In: The European control conference, pp 4320–4325
2. Zhang H, Osterlind F, Soldati P, Voigt T, Johansson M (2015) Time-optimal convergecast with separated packet copying: scheduling policies and performance. Trans Veh Technol 64:793–803
3. Saifullah A, Xu Y, Lu C, Chen Y (2011) Priority assignment for real-time flows in WirelessHART networks. In: The Euromicro conference on real-time systems (ECRTS), pp 35–44
4. Burns A, Davis RI (2022) Mixed criticality systems - a review. https://www-users.cs.york.ac.uk/burns/review.pdf
5. Baruah S, Vestal S (2008) Schedulability analysis of sporadic tasks with multiple criticality specifications. In: The Euromicro conference on real-time systems (ECRTS), pp 147–155
6. Burns A, Harbin J, Indrusiak LS (2014) A wormhole NoC protocol for mixed criticality systems. In: The real-time systems symposium (RTSS). IEEE, Piscataway, pp 184–195
7. Tobuschat S, Axer P, Ernst R, Diemer J (2013) IDAMC: a NoC for mixed criticality systems. In: The international conference on embedded and real-time computing systems and applications (RTCSA). IEEE, Piscataway, pp 149–156
8. Saifullah A, Xu Y, Lu C, Chen Y (2010) Real-time scheduling for WirelessHART networks. In: Real-time systems symposium (RTSS). IEEE, Piscataway, pp 150–159
9. Zhang HB, Soldati P, Johansson M (2009) Optimal link scheduling and channel assignment for convergecast in linear WirelessHART networks. In: The international symposium on modeling and optimization in mobile, ad hoc, and wireless networks, pp 1–8
10. Chipare O, Lu CY, Roman GC (2013) Real-time query scheduling for wireless sensor networks. Trans Comput 62(9):1850–1865
11. Carvajal G, Fischmeister S (2013) An open platform for mixed-criticality real-time ethernet. In: The design, automation and test in europe conference and exhibition. IEEE, Piscataway, pp 153–156
12. Koubâa A, Alves M, Tovar E, Cunha A (2008) An implicit GTS allocation mechanism in IEEE 802.15.4 for time-sensitive wireless sensor networks: theory and practice. Real-Time Syst 39:169–204
13. Zhan Y, Xia Y, Anwar M (2016) GTS size adaptation algorithm for IEEE 802.15.4 wireless networks. Ad Hoc Netw 37:486–498
14. Huang HM, Gill C, Lu CY (2014) Implementation and evaluation of mixed-criticality scheduling approaches for sporadic tasks. Trans Embed Comput Syst 13(4):1–25
15. Vestal S (2007) Preemptive scheduling of multi-criticality systems with varying degrees of execution time assurance. In: Real-time systems symposium (RTSS). IEEE, Piscataway, pp 239–243
16. Burns A, Fleming T, Baruah S (2015) Cyclic executives, multi-core platforms and mixed criticality applications. In: Euromicro conference on real-time systems (ECRTS). IEEE, Piscataway, pp 3–12
17. Lee J, Phan KM, Gu XZ, Lee JY, Easwaran A, Shin I, Lee I (2014) Mc-fluid: fluid model-based mixed-criticality scheduling on multiprocessors. In: Real-time systems symposium (RTSS). IEEE, Piscataway, pp 41–52
18. Burns A, Davis RI (2013) Mixed criticality on controller area network. In: The Euromicro conference on real-time systems (ECRTS). IEEE, Piscataway, pp 125–134

19. Cros O, Fauberteau F, George L, Li X (2014) Mixed-criticality over switched ethernet networks. In: The workshop on mixed criticality for industrial systems, pp 138–143
20. Addisu A, George L, Sciandra V, Agueh M (2013) Mixed criticality scheduling applied to JPEG2000 video streaming over wireless multimedia sensor networks. In: The 1st workshop on mixed criticality systems. IEEE, Piscataway, pp 1–6
21. Shen W, Zhang T, Barac F, Gidlund M (2014) PriorityMAC: a priority-enhanced MAC protocol for critical traffic in industrial wireless sensor and actuator networks. Trans Ind Inf 10:824–835
22. Hussain SW, Khan T, Zaidi SH (2006) Latency and energy efficient MAC (LEEMAC) protocol for event critical applications in WSNs. In: The international symposium on collaborative technologies and systems, pp 370–378
23. Liang W, Zhang XL, Xiao Y, Wang F, Zeng P, Yu HB (2011) Survey and experiments of WIA-PA specification of industrial wireless network. Wirel Commun Mob Comput 11(8):1197–1212
24. IEC (2016) Industrial networks - wireless communication network and communication profiles - WirelessHARTTM. International Electrotechnical Commission
25. Li B, Nie LS, Wu CJ, Gonzalez H, Lu CY (2015) Incorporating emergency alarms in reliable wireless process control. In: ACM/IEEE international conference on cyber-physical systems. ACM/IEEE, Piscataway, pp 218–227
26. Baruah S, Bonifaci W, DAngelo G, Li HH, Marchetti-Spaccamela A, Megow N, Stougie L (2012) Scheduling real-time mixed-criticality jobs. Trans Comput 61(8):1140–1152
27. De Moura L, Bjorner N (2008) Z3: an efficient SMT solver. In: Tools and algorithms for the construction and analysis of systems. Springer, Berlin, pp 337–340
28. Dutertre B, De Moura L (2006) Tools and algorithms for the construction and analysis of systems. In: Satisfiability modulo theories competition, pp 1–3
29. Pellizzoni R, Paryab N, Yoon MK, Bak S, Mohan S, Bobba RB (2015) A generalized model for preventing information leakage in hard real-time systems. In: Real-time and embedded technology and applications symposium (RTAS). IEEE, Piscataway, pp 271–282
30. Guan N, Stigge M, Wang Y, Yu G (2009) New response time bounds for fixed priority multiprocessor scheduling. In: Real-time systems symposium. IEEE, Piscataway, pp 387–397
31. Saifullah A, Xu Y, Lu CY, Chen YX (2011) End-to-end delay analysis for fixed priority scheduling in WirelessHART networks. In: Real-time and embedded technology and applications symposium. pp 13–22
32. Joseph M, Pandya P (1986) Finding response times in a real-time system. Comput J 29(5):390–395
33. Camilo T, Silva JS, Rodrigues A, Boavida F (2007) Gensen: a topology generator for real wireless sensor networks deployment. In: The international conference on software technologies for embedded and ubiquitous systems. Springer, Berlin, pp 436–445
34. Bini E, Buttazzo CC (2005) Measuring the performance of schedulability tests. Real-Time Syst 30(1):129–154

Chapter 5
Mixed-Criticality Scheduling with Multiple Radio Interfaces

Abstract In this chapter, we introduce the nodes with multiple radio interfaces (MRI) into mixed-criticality industrial wireless networks. When an error occurs or transmission demand changes, the MRI nodes can switch their transmission mode, changing to a high-criticality configuration to meet the system's new demand. Hence, we first propose a heterogeneous MRI system model. Based on this model, we propose a Slot Analyzing Algorithm (SAA) that guarantees system schedulability by reallocating slots for each node after replacing conflict nodes with MRI nodes. By considering both system schedulability and cost, SAA also reduces the number of MRI nodes. Then, we propose a Priority Inversion Algorithm (PIA) that improves the schedulability by adjusting slot allocations before replacing conflict nodes with MRI nodes. By reducing the use of MRI nodes, PIA achieves better performance than SAA when the system is in the high-criticality mode.

5.1 Background

Due to the high real-time and reliability requirements of industrial systems, traditional cyber-physical systems cannot be applied directly [1–4]. To guarantee the requirements of ICPSs, we introduce multiple radio interface (MRI) nodes into ICPSs. A traditional network node is usually equipped with only one antenna, while an MRI node with two antennas, can both receive and send packets simultaneously [5–7]. By replacing conflict nodes with MRI nodes, multiple flows can transmit without delays. If we were to replace all the conflict nodes with MRI nodes, that would eliminate transmission conflicts in the system; consequently, when the number of MRI nodes is sufficient, we can guarantee system schedulability. However, considering the energy consumption and cost of MRI nodes, it is advantageous to be able to guarantee the network schedulability using fewer MRI nodes [8–12]. Hence, there are two main problems in mixed-criticality industrial wireless networks, (1) reducing the number of MRI nodes required to guarantee the network schedulability and (2) analyzing the network schedulability under different criticality modes. To address these problems, we propose two algorithms, a Slot Analyzing Algorithm (SAA) and a Priority Inversion Algorithm (PIA), to

© The Author(s) 2023
X. Jin et al., *Mixed-criticality Industrial Wireless Networks*, Wireless Networks,
https://doi.org/10.1007/978-981-19-8922-3_5

improve the network schedulability. SAA first allocates slots without considering transmission conflicts and replaces conflict nodes with MRI nodes to guarantee future schedulability in low-criticality mode. In contrast, PIA first optimizes slot allocation to reduce the number of transmission conflicts; then, we replace conflict nodes with MRI nodes only when the system cannot be scheduled. The contents of this chapter are as follows:

1. We first propose an algorithm to obtain the candidate node set (CNSA), which indicates the key nodes that affect system performance, and then propose SAA to improve system schedulability. SAA can guarantee that a system can be scheduled when it satisfies Theorem 5.1. Furthermore, SAA also reduces the number of MRI nodes through slot analysis.
2. We propose PIA to reduce the number of transmission conflicts before replacing the overlap nodes with MRI nodes. When a flow cannot be scheduled, priority inversion occurs when it satisfies Theorem 5.3.
3. We analyze system schedulability in high-criticality mode and prove that PIA achieves better schedulability than SAA in Theorem 5.4.
4. Simulation results show that our algorithms can improve the schedulability of industrial wireless networks with MRI nodes and that PIA achieves a better performance than SAA in the high-criticality mode.

5.2   System Model

We consider a heterogeneous network consisting of field devices (both MRI nodes and traditional sensor nodes), one centralized controller and one gateway. In this section, firstly, we propose a network model, and then we introduce MRI nodes into mixed-criticality networks. Finally, we apply a fixed priority (FP) scheduling scheme in industrial wireless networks.

5.2.1   Network Model

The system model is based on WIA-PA [13]. There are several features as follows: (1) Time Division Multiple Access (TDMA); (2) Route and Spectrum Diversity and (3) Handling Internal Interference. The number of channels is denoted as m, and there are N fixed nodes in our system.

Industrial systems have higher requirements of real-time performance and reliability. Here, we introduce MRI nodes into industrial systems. MRI nodes are promising for wireless transmissions because they can improve the average user spectral efficiency [14]. By mounting multiple antennas on a single sensor node, the node can receive and transmit simultaneously (or improve the packet acceptance ratio when all the antennas work on the same channel).

Fig. 5.1 An example

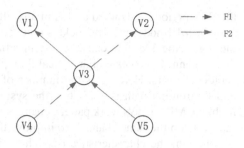

Hence, our model includes two types of sensor nodes: traditional nodes and MRI nodes. Each traditional node in our system is equipped with a half-duplex omni-directional radio transceiver whose status can alternate between transmitting and receiving. In contrast, an MRI node has two working modes: (1) when the antennas work on different channels, the node can both receive and transmit packets from several paths (the transmission capacity of an MRI node depends on the number of antennas it has); and (2) when the antennas work on the same channel, the node can reduce signal interference and improve the acceptance ratio because several antennas receive a packet from the same flow. Each packet in our system is transmitted through the network under source routing.

Figure 5.1 shows an example of an MRI node in the first mode (where antennas work on different channels) to address a transmission conflict in which F_1 and F_2 send packets to the same destination simultaneously. We introduce MRI to solve this issue. The flows can be scheduled by replacing $V3$ with an MRI node. MRI nodes can also be used as a method to increase the packet acceptance ratio when two or several antennas receive packets on the same channel. For the second mode of an MRI node (where the antennas work on the same channel), when the links around $V3$ have poor signal quality (such as from co-channel interference, intermodulation interference, spurious emissions or adjacent-channel interference), node $V3$ can switch its antenna receiving frequencies to the same channel to improve the acceptance ratio. However, considering that the power consumption and cost of MRI nodes is considerably higher than the cost of a normal node, we cannot deploy MRI nodes throughout the entire system.

5.2.2 Mixed-Criticality System

Due to the introduction of MRI nodes, our mixed-criticality system model differs from other models [15–19]. The flow is a periodic end-to-end communication between a source and its destination. There are n flows in our system, denoted by $F = \{F_1, F_2, \ldots, F_n\}$. F_i is characterized by $< t_i, d_i, \xi, c_i, p_i >$, a number between $1 \leq i \leq n$, where the period is t_i, the deadline is d_i, the criticality level is ξ (we focus on a dual-criticality system $\{LO, HI\}$) in which $\xi = 2$, meaning the system has two critical levels. The superframe as the lowest common multiple of the flow periods is denoted as T. Initially, the system works at a low critical level, and the MRI nodes receive packets on different channels. The system switches to the high critical level to enhance system reliability, and the MRI nodes reconfigure their antennas to receive high critical flow packets on the same channels when the accident occurs. Here, c_i is the number of hops from a source to a destination, and the routing path is p_i. When the system critical level switches from low to high, the MRI node's work pattern changes from receiving on different channels to receiving on the same channels to improve the schedulability of high critical flows. In addition, the characteristics of high-criticality flows switch to high-criticality mode. In this chapter, the system reduces the sampling period of a high-criticality flow to $\aleph_i t_i$ to improve the sampling rate of high critical flow, where $\aleph_i$ is a value that satisfies $\frac{c_i}{t_i} \leq \aleph_i < 1$. To simplify the calculation we assume that $d_i = t_i$. Hence, $\aleph_i$ satisfies $\frac{c_i}{d_i} \leq \aleph_i < 1$. We model the duration of the mode switch as γ, which is used to calculate the delay of packets delivered during the mode change.

In the beginning, packets are transmitted in low critical mode. To provide more resources, each MRI node's antennas work on different channels. The system switches when the accident occurs. To enhance system reliability in high-criticality mode, each MRI node's antennas are reconfigured to work on the same channel.

5.2.3 Fixed Priority Scheduling

We provide a summary of fixed priority scheduling to analyze the schedulability of systems in this subsection. FP scheduling is a commonly adopted scheme in practice for cyber-physical systems and real-time CPU scheduling [20]. Each job priority is pre-allocated by the network controller, and transmissions are scheduled based on this priority.

We assume that the priority of each flow is the same as its number. That is, F_1 has the highest priority, and F_n has the lowest priority in the system. There are two types of delays in industrial wireless sensor systems: (1) Channel contention and (2) transmission conflicts, the definitions can be obtained in [21].

We define a system as schedulable when all the flows in a system can be scheduled (reach the destination before its deadline). Then the definition of network schedulability is whether or not all flows in a network are schedulable. It is worth noting that when the system is in high-criticality mode, we no longer focus on the schedulability of low-criticality flows. When we repeat Z experiments, and only z experiments the emergency flow can be scheduled, then, the schedulability ratio is $\frac{z}{Z}$.

5.3 Problem Statement

Given the flow set F, a wireless network, and the FP scheduling policy, our objective is to use MRI nodes to improve the schedulability of mixed-criticality industrial networks. We first analyze the schedulability of mixed-criticality industrial wireless networks. Initially, the system is working in low-criticality mode. To improve system schedulability, we introduce MRI technology to provide more resources (the resources such as slots and channels are increased since MRI nodes can receive and send packets simultaneously). Hence, we replace nodes at several intersections with MRI nodes and propose a Priority Inversion Algorithm to guarantee that the system can be scheduled. However, when the accident occurs, we must guarantee high-criticality flows schedulability; consequently, the system switches to high-criticality mode, and the transmission mode of MRI nodes changes. We then analyze the schedulability of our method to evaluate the quality of PIA. The challenges in this situation are listed below.

1. When the system is deployed, we can easily determine flows that miss their deadlines. However, each flow's schedulability is interrelated with others. Thus, how should we decide which nodes should be replaced with MRI nodes?
2. As described in the previous section, the power consumption and cost of an MRI node are much higher than those of a normal node. Therefore, it is unreasonable and not cost-effective to deploy many MRI nodes when the system can be scheduled. Hence, the problem of determining the smallest number of MRI nodes that can meet the system requirements in low-criticality mode is another challenge.
3. The MRI nodes' transmission mode changes when the system switches to high-criticality mode. Therefore, the scheduling algorithm also needs to consider schedulability in high-criticality mode.
4. We need to analyze the system's schedulability under our proposed method.

5.4 Scheduling Algorithms

In this section, we first study the issue of how to improve system schedulability with a small number of MRI nodes. First, we identify the nodes that may be replaced with MRI nodes to define candidate nodes as follows:

Definition 5.1 (Candidate Node) We define a candidate node as a node at which transmission conflicts (intersection or overlap nodes) can occur. As Fig. 5.1 shows, the paths of F_1 and F_2 intersect at $V3$; consequently, a transmission conflict may occur at this node. Thus, $V3$ is a candidate node.

Flows may conflict when they have path overlaps. To facilitate candidate node identification, we assume that, at most, one part of a path overlaps between two flows. As Fig. 5.2 shows, two periodic flows transmit in a network that conflict at

Fig. 5.2 The analysis of conflicts. (**a**) Routing. (**b**) Superframe. (**c**) Scheduling

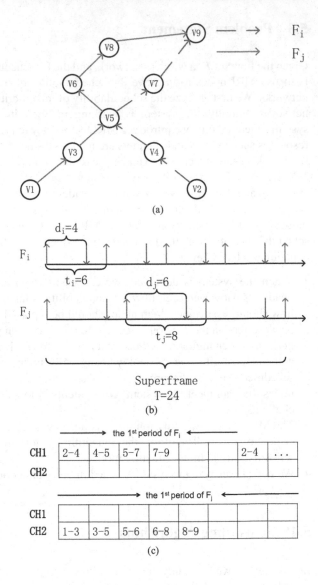

the second and third slots in the first period. There are two methods for addressing transmission conflicts. The first method is to reallocate the slots for each node after replacing the conflict nodes with MRI nodes (as Fig. 5.2 show, the system can be scheduled when we replace node $V5$ with an MRI node, where CH is the abbreviation of Channel); the other method is to adjust slot allocation, as shown in Fig. 5.3. By rationally allocating each flow's transmission slots, we can reduce the number of MRI nodes and the signal interference caused by the coexistence of multiple channels, which can also improve the schedulability when the wireless network is in high-criticality mode. However, this is unsuitable for a system that

Fig. 5.3 An example for un-schedulable conflict

	⟶ the 1st period of F_i ⟵							
CH1	2–4			4–5	5–7	7–9	2–4	. . .
CH2								

			⟶ the 1st period of F_j ⟵					
CH1								
CH2	1–3	3–5	5–6	6–8	8–9			

cannot be scheduled (in this example, the system cannot be scheduled if the deadline of F_i is 4).

Therefore, for a system in which each link's transmission slot has been allocated, we propose the slot analyzing algorithm to improve system schedulability under a traditional FP policy. When the system cannot be scheduled by MRI nodes, SAA re-allocates system resources to guarantee system schedulability (increase the transmission speed of unscheduled data flows). By considering the characteristics of mixed-criticality systems and MRI nodes, we then propose the PIA algorithm, which adjusts slot allocations before replacing intersection nodes with MRI nodes (the second method to address transmission conflict). By optimizing slot allocation, PIA can guarantee system schedulability with fewer MRI nodes. To guarantee the schedulability of the network, PIA will also re-allocate slots when the system cannot be scheduled. It is important to note that initially both the SAA and PIA algorithms work in low-criticality mode. We will analyze the schedulability of these two algorithms later.

5.4.1 Finding Candidate Nodes

A network consists of numerous nodes. In this subsection, we study how to select candidate nodes to guarantee the system can be scheduled. Transmission conflicts occur at the path overlaps of flows. As Fig. 5.4 shows, there are two types of overlaps (without considering the flow's direction).

Fig. 5.4 Transmission conflict

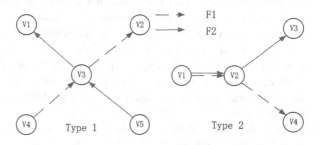

Lemma 5.1 *When flow paths have an overlapping region, the overlap nodes are candidates for MRI nodes. We denote the candidate node set as Λ, and the node set on each flow's path can be denoted as λ_i.*

Proof Transmission conflict can obviously only occur at an overlapping region. We can solve this issue by MRI nodes. We can account for the conflicts caused by the first type as shown in Fig. 5.1. For the transmission conflict caused by the second type, as Fig. 5.4 shown, there are two nodes ($V1$ and $V2$) on both F_1 and F_2 involved in the second type of overlap. When this type of transmission conflict occurs, we can improve the schedulability by replacing both $V1$ and $V2$ with MRI nodes. Hence, when flow paths have an overlapping region, the overlap nodes are candidates for MRI nodes.   □

Hence, we propose a Candidate Node Searching Algorithm (CNSA) to search for the set of candidate nodes as follows:

Algorithm 5.1 Candidate node searching algorithm

Input: F;
Output: the candidate node set $\Lambda = \{\lambda_i\}, i \in F$;
 1: **for** each flow F_i **do**
 2:    **if** node n_k is the overlap node **then**
 3:        n_k join Λ;
 4:    **end if**
 5: **end for**
 6: **return** Λ;

We search the candidate nodes by traversing the flow paths of the entire system. Nodes on more than one flow path are added to the candidate node set Λ. The time complexity of CNSA is $O(F^2)$.

5.4.2 Slot Analyzing Algorithm

After obtaining the candidate node set, we reduce the number of nodes in this set to reduce the system's cost. Because not all the candidate nodes can experience transmission conflicts, we propose the slot analyzing algorithm to analyze the schedulability of the network under a traditional FP policy in one superframe (a superframe is the lowest common multiple of t_i, $i \in F$). SAA improves system schedulability by replacing some of the nodes in Λ with MRI nodes.

Since SAA is used after the CNSA, we do not consider the slot allocations for each flow; we just replace conflict nodes with MRI nodes when a transmission conflict occurs.

When the system cannot be scheduled, we need to improve the schedulability using MRI nodes. Because MRI nodes can both receive and send packets in a single time slot, we first allocate slots for each node without considering transmission conflicts (the network controller allocates the channel for each transmission using the FP scheduling policy). Then, we replace conflict nodes with MRI nodes to guarantee system schedulability. Obviously, it is both unnecessary and not cost-effective to replace all the nodes in the candidate node set with MRI nodes. Therefore, SAA reduces the number of MRI nodes in Λ through slot analysis. The pseudo code of the SAA algorithm is as follows:

Algorithm 5.2 Slot analyzing algorithm

Input: the characters for each flow F_i; the candidate node set $\Lambda = \{\lambda_i\}$, $i \in F$;
Output: the schedulability of the network;
1: reallocate the slot for each node.
2: **for** each flow i **do**
3:    **if** the flow cannot be scheduled **then**
4:        find the intersection nodes and reallocate slots for F_i without considering transmission conflicts.
5:    **else**
6:        retain the original allocation.
7:    **end if**
8: **end for**
9: **for** each node $\lambda_i \in \Lambda$ **do**
10:    **if** λ_i has two or more than two transmissions in the same time slot in one superframe **then**
11:        λ_i needs to be replaced;
12:    **end if**
13: **end for**
14: **if** the system also cannot be scheduled **then**
15:    **for** each unscheduled flow **do**
16:        enhance the unscheduled flow's transmission speed;
17:        **if** λ_i has two or more than two transmissions in the same time slot in one superframe **then**
18:            λ_i needs to be replaced;
19:        **end if**
20:    **end for**
21: **end if**
22: update λ_i and $\Lambda = \{\lambda_i\}$;
23: **return** Λ;

We reallocate the transmission time slots for each node by the schedulability of each flow (lines 1–8). If the flow cannot be scheduled, we find the intersection nodes and reallocate the slots for this flow without considering transmission conflicts. Otherwise, we retain the original allocation. Then, we analyze the transmission slot for each node in Λ (lines 9–23). When a node in Λ has more than one transmission at the same time slot, that means a transmission conflict occurs at this node, and we need to replace the node with an MRI node (lines 9–13). When the system also cannot be scheduled, SAA then re-allocates slots to accelerate the transmission speed of unscheduled flows until the flows can be scheduled (lines 14–21). Finally,

SAA updates λ_i and returns Λ—the set of nodes that need to be replaced (lines 22–23). The theorem is as follows:

Theorem 5.1 *A network can be scheduled with SAA when the number of channels is no less than the number of flows ($m \geq n$).*

Proof When k flows conflict at node A, we denote the flow with the highest priority as F_1, and the flow with the lowest priority as F_k. F_1 transmits first and cannot be delayed at node A. The other flows must wait on F_1,consequently, this generates delay. When the number of channels is no less than the number of flows, no delay is caused by transmission conflicts. Hence, flow $F_i, i \in k$ will miss its deadline and cannot be scheduled when $c_i + del_i > d_i$, where del is the delay slots. Using SAA, we can eliminate the delays caused by transmission conflicts. k flows can transmit simultaneously when $m > n$ because the number of hops in each flow's transmission c is no larger than its deadline d. Thus, we can guarantee all the flows can be scheduled and, therefore, the system can be scheduled using SAA when the number of channels is no less than the number of flows. $\Box$

5.4.3 Priority Inversion Algorithm

In this subsection, we study how to reduce the number of MRI nodes by the second method, the PIA scheduling algorithm, which performs optimal allocation of resources before the system runs. Through optimal allocation, PIA can decrease the number of transmission conflicts at intersection nodes and further reduce the number of required MRI nodes. As Fig. 5.5 shows, $V5$ does not need to be replaced by an MRI node when $d_i = 6$ in Fig. 5.2a.

Initially, PIA allocates slots based on the traditional FP policy. If the system cannot be scheduled, the network controller obtains each flow's arrival time at its destination, which can be denoted as r_i. Then, we can obtain the following theorem,

Theorem 5.2 *When F_i and F_j conflict at node set $V = V_1, V_2 \ldots, V_h$ ($i < j$), we denote the length of the overlap as $Len(ij)$, and its conflict delay is $D(ij)$. The schedulability of F_i is unaffected by the priority inversion of F_i and F_j when it satisfies:(1) $r_i + D(ij) \leq d_i$;(2) V is the last part of the overlapping nodes of F_i.*

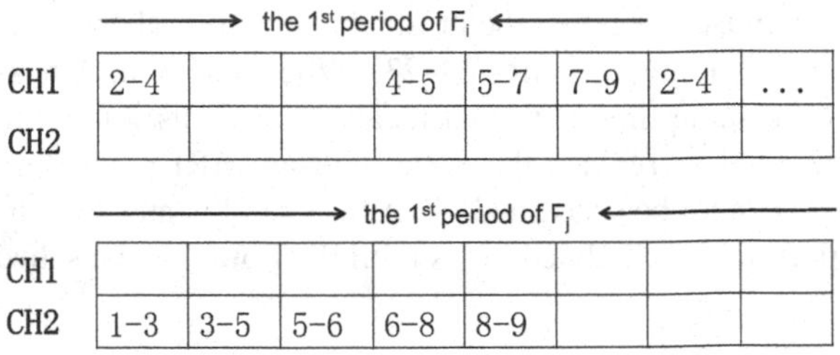

Fig. 5.5 Transmission conflict

the 1st period of F_i

CH1	2–4			4–5	5–7	7–9	2–4	...
CH2								

the 1st period of F_j

CH1								
CH2	1–3	3–5	5–6	6–8	8–9			

Proof The priority of F_i is higher than that of F_j based on the assumption that $i < j$. When transmission conflict occurs at node set $V = V_1, V_2 \ldots, V_h$, F_j must be delayed by F_i. We can accelerate the transmission speed by transmitting F_j first when the schedulability of F_i does not be changed. We introduce the result proposed by Saifullah in [22], in which the upper bound $\Delta(i, j)$ of the delay that F_j can experience from an instance of F_i is

$$\Delta(ij) = \sum_{k=1}^{\delta(ij)} Len_k(ij) - \sum_{k'=1}^{\delta'(ij)} (Len_{k'}(ij) - 3), \tag{5.1}$$

where $\delta(ij)$ is the number of path overlaps, $Len_k(ij)$ is length of the kth overlap on F_i and F_j, $\delta'(ij)$ is the number of path overlaps larger than 3. Hence, the problem is transformed into one of calculating the delay F_i must bear while waiting for F_j to transmit. By simplifying Eq. (5.2), we can obtain the delay in V as follows:

$$D(ij) = \begin{cases} Len(ij), & Len(ij) < 4, \\ Len(ij) - 3, & Len(ij) \geq 4. \end{cases} \tag{5.2}$$

When V is the last part of the overlap nodes of F_i, no other factors can cause additional delay to F_i. Hence, the schedulability of F_i is unaffected by the priority inversion of F_i and F_j when it satisfies:(1) $r_i + D(ij) \leq d_i$; and (2) V is the last part of the overlap nodes of F_i.   □

Based on Theorem 5.2, PIA accelerates the unscheduled flow until it can be scheduled or no longer influences the other flow's schedulability. However, Theorem 5.2 is not suitable for all overlaps. When F_i involves several parts of overlaps, we can only guarantee the acceleration of the low priority flow on the last overlap. For the flow's intersections with F_i in the other parts of the overlaps, we define an Effective Overlap Region(EOR) as follows:

Definition 5.2 (Effective Overlap Region) When F_i and F_j conflict in node set $V = V_1, V_2 \ldots, V_h$ ($i < j$), the overall node set V is the effective overlap region. If F_i and F_j have an overlap but do not conflict, then V is not an effective overlap region.

Hence, Theorem 5.2 can be extended as follows:

Theorem 5.3 When F_i and F_j conflict at node set $V = V_1, V_2 \ldots, V_h$ ($i < j$), the schedulability of F_i is unaffected by the priority inversion of F_i and F_j when it satisfies:(1) $r_i + D(ij) \leq d_i$; (2) it does not change the character of non-EORs after V; and (3) V is the last EOR of F_i.

Proof Based on Theorem 5.2 we extend priority inversion to the last part of the EOR of F_i. When the operation satisfies (1) $r_i + D(ij) \leq d_i$; (2) it does not change the character of non-EORs after V; and (3) V is the last part of the EOR of F_i. Then, the schedulability of F_i and its intersection flows are unaffected by the wait for F_j

to transmit. Hence, when F_i and F_j conflict at node set $V = V_1, V_2 \ldots, V_h$ $(i < j)$, the schedulability of F_i is unaffected by the priority inversion of F_i and F_j when it satisfies:(1) $r_i + D(ij) \leq d_i$; (2) it does not change the character of non-EORs after V; and (3) V is the last part of the EOR of F_i. $\quad\square$

To guarantee the second constraint condition in Theorem 5.3, we denote the number of overlaps after the last EOR in F_i as o_i. For any flow k, when F_k and F_i have a path overlap after the last EOR in V, we denote this overlap path as o_{ik}. Hence, the fact that the second condition in Theorem 5.3 can be translated into the priority inversion operation does not affect the schedulability of F_i and F_k. Obviously, F_k and F_i cannot conflict when the packets generated by the two flows arrive such that o_{ik} satisfies

$$\|r_i^{o_{ik}} - r_k^{o_{ik}}\| > Dij, \tag{5.3}$$

where $r_i^{o_{ik}}$ is the time at which the packet generated by F_i reaches o_{ik} (it is easy to obtain o_{ik} from the network controller). When the arrival time difference of F_i and F_k is larger than Dij, o_{ik} is not an EOR. By updating $r_i^{o_{ik}}$, $i \in F$, we can determine whether to invert the priority of F_i and F_j.

Based on the above discussion, PIA can reduce the number of transmission conflicts before adding MRI nodes to the system. Furthermore, PIA can also improve the schedulability of networks under a high-criticality mode. When the system cannot be scheduled under PIA, the nodes at flow intersections will be replaced by MRI nodes to guarantee the schedulability of the network. The pseudo code of PIA is as follows,

---

Algorithm 5.3 Priority inversion algorithm

Input: the characters for each flow F_i; the candidate node set $\Lambda = \{\lambda_i\}, i \in F$.
Output: reduce the number of transmission conflicts by optimal slot allocations;
1: reallocation slots for each node by FP.
2: the number of MRI nodes $\tau = 0$.
3: **while** the system cannot be scheduled **do**
4:    **for** each flow i **do**
5:       obtain each flow's o_i, Len and the corresponding $o_{ik}, i, k \in F$;
6:    **end for**
7:    **for** each unscheduled flow j **do**
8:       **if** flow i satisfies Theorem 5.3 **then**
9:          invert the priority between F_i and F_j;
10:      **else**
11:         replace the overlap nods with MRI nodes;
12:         $\tau + +$;
13:      **end if**
14:    **end for**
15: **end while**
16: **return** τ;

PIA first allocates slots under the FP policy (line 1) and initializes the number of MRI nodes (line 2). Then, it judges whether the system can be scheduled or not. If the system can be scheduled, it returns τ, otherwise, PIA accelerates the transmission speed of unscheduled flows by Theorem 5.3. When the flows do not satisfy the conditions, PIA replaces the overlap nodes with MRI nodes, similar to SAA, and increments τ by one (lines 3–16).

When the system performs the priority inversion operation at the last EOR, the conflict is removed. Hence, that part of the overlap is no longer an EOR. The system repeats this process until either the system can be scheduled or no flow satisfies Theorem 5.3.

5.4.4 Algorithm Analysis in High-Criticality Mode

In this subsection, we analyze the schedulability of SAA and PIA. Since both SAA and PIA can be scheduled by adding MRI nodes in low-criticality mode, we analyze only the schedulability in high-criticality mode.

Theorem 5.4 *PIA has a higher schedulability than SAA when the system is in high-criticality mode.*

Proof When the system switches to high-criticality mode, low-criticality flows are abandoned and the high criticality flow period changes to $\aleph_i t_i$, $\frac{c_i}{t_i} \leq \aleph_i < 1$. Because $t_i = d_i$, the flow must arrive at its destination before $\aleph_i d_i$. We analyze the schedulability by a discussion of classification. The flows are classified into two categories: (1) those in which the flow does not overlap other high-criticality flows and (2) those in which the flow overlaps other high-criticality flows. For F_i in the first category, the schedulability of both SAA and PIA are identical. When a flow is not blocked by other flows it can be scheduled when

$$r_i \leq \aleph_i d_i. \tag{5.4}$$

For F_j in the second category, F_j is delayed by other flows in $D(j)$ slots. Then, F_j can be scheduled when

$$r_j + D(j) \leq \aleph_j d_j. \tag{5.5}$$

We denote the number of MRI nodes in SAA and PIA as τ_j^{SAA} and τ_j^{PIA}, respectively. Because PIA reduces the number of transmission conflicts before replacing intersection nodes with MRI nodes, $\tau_j^{SAA} \geq \tau_j^{PIA}$. By Theorem 5.2, we can obtain

$$D(j) = \sum_{k=1}^{\tau_j} D(ij), \tag{5.6}$$

when F_i and F_j conflict, $D(ij) = \begin{cases} Len(ij), & Len(ij) < 4 \\ (Len(ij) - 3), & Len(ij) \geq 4 \end{cases}$, which is

determined only by the length of the overlap. Then, we can obtain the relationship between SAA and PIA as follows:

$$r_j^{PIA} + D(j)^{PIA} \leq r_j^{SAA} + D(j)^{SAA}. \tag{5.7}$$

Hence, PIA has a higher schedulability than SAA when the system is in high-criticality mode.   □

5.5 Performance Evaluations

We evaluate the performance of our proposed methods by experiments. We compare our approaches with the traditional FP algorithm without MRI nodes. We compare both the acceptance rate and the number of MRI nodes for each criticality mode. We use the acceptance rate to represent the schedulability of a network. When all flows can be scheduled, the acceptance rate is 1; otherwise, it is 0. To control the workload of the entire system, the simulations use the UUniFast algorithm, which can make the flows neither pessimistic nor optimistic for the analysis [23]. All algorithms are implemented in C language. These programs run on a Windows machine with 3.2 GHz CPU and 8 GB memory. Some simulation parameters are summarized in Table 5.1.

5.5.1 Low-Criticality Mode

We first compare the performances of the algorithms in low-critical mode. As Fig. 5.6(a) shown, the system acceptance rate is decreased by the FP scheduling policy (n=15, N=50, m=16), which occurs because the idle resources decrease as the system utilization increases. The latency tolerance of a packet is reduced along

Table 5.1 Simulation parameters

Parameter	Description
n	The number of flows
N	The number of nodes
m	The number of channels
U	Network utilization
u_i	Flow i's utilization
t	The period of flow
d	The deadline of flow
c	The number of transmission hops

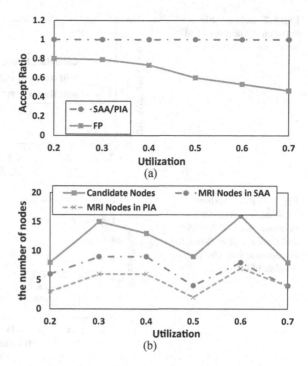

Fig. 5.6 Relationship between the acceptance rate/the number of MRI nodes and system utilization. (**a**) Acceptance rate. (**b**) The number of nodes

with the idle resources. The network can be scheduled under both SAA and PIA in any situation by increasing the number of MRI nodes when $m \geq n$. Figure 5.6b shows the relationship between the number of MRI nodes and utilization. Obviously, both SAA and PIA can reduce the number of MRI nodes under the premise that the system is schedulable. Because PIA optimizes slot allocation before the node replacement operation, PIA uses fewer MRI nodes than SAA. However, none of the curves have an obvious tendency because system utilization involves not only the flow period (t) but also the number of transmission hops (c). We need to regenerate the transmission path for each flow to satisfy the system utilization requirements. Hence, the number of candidate nodes goes up and down. Because an MRI node is selected from the candidate node set, the number of MRI nodes is always less than the number of candidate nodes.

We repeat this simulation for the situation in which all flows can be scheduled as shown in Fig. 5.7. The number of candidate nodes is fixed when we increase system utilization by only adjusting the flow period (because there is only one test in this simulation, and the flow transmission paths do not vary). Initially, no MRI nodes exist in the system because they can be scheduled without any MRI nodes. However, when we increase the system utilization, transmission conflict occurs. To guarantee the schedulability of the system, we need to add MRI nodes to the system. In addition, the number of MRI nodes required by PIA is always less than the number required by SAA. This occurs because by optimizing slot allocation,

Fig. 5.7 Relationship
between the number of MRI
nodes and system utilization

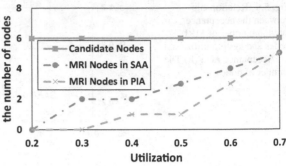

Fig. 5.8 Relationship
between the acceptance
rate/number of MRI nodes
and the number of flows. (a)
Acceptance rate. (b) The
number of nodes

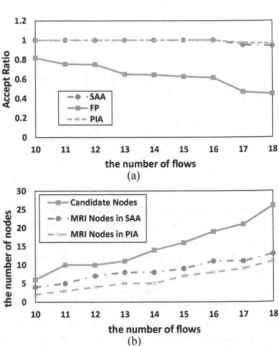

PIA can reduce the number of transmission conflicts. Hence, the number of MRI nodes required by PIA is always less than the number required by SAA.

The relationship between the acceptance rate/number of MRI nodes and the number of flows is shown in Fig. 5.8 (U=0.3, N=50, m=16). The acceptance rate is reduced along with the number of flows under the FP scheduling policy. The acceptance rates of SAA and PIA are reduced when the number of flows becomes larger than the number of channels. This result occurs because when $m < n$, delays are caused by channel contention. PIA achieves a better performance than SAA when $n > 16$ because it uses fewer MRI nodes. In addition, both the number of candidate nodes and the number of MRI nodes increase as the number of flows increases. This result occurs because the number of intersections increases when

Fig. 5.9 Relationship between the acceptance rate/number of MRI nodes and the number of flows. (a) Acceptance rate. (b) The number of nodes

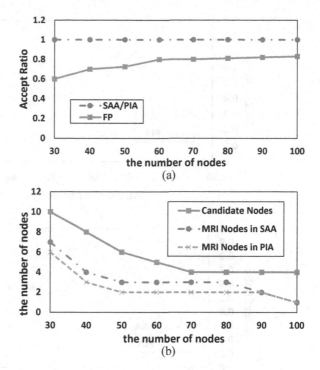

the number of flows increases, causing more transmission conflicts in the system. Thus, the system requires more MRI nodes to guarantee its schedulability, and the number of MRI nodes always satisfies $\tau^{FP} > \tau^{SAA} > \tau^{PIA}$.

Figure 5.9 shows the relationship between the acceptance rate/number of MRI nodes and the number of nodes (U=0.3, n=15, m=16). All these results indicate that SAA and PIA can guarantee the schedulability of the system. The number of MRI nodes in both SAA and PIA is no larger than the number of candidate nodes regardless of the conditions. In addition, when the number of nodes increases to 90, the number of MRI nodes in PIA and SAA is identical because PIA transmutes into SAA when it cannot resolve the conflict.

5.5.2 High-Criticality Mode

When the system switches to high-critical mode, the number of MRI nodes is no longer our main concern (the number of MRI nodes is the same as in low-critical mode). Instead, we are more concerned with the performances of SAA and PIA in high-critical mode.

Figure 5.10 shows the relationship between the system acceptance rate and system parameters such as the number of nodes and the number of flows. In these

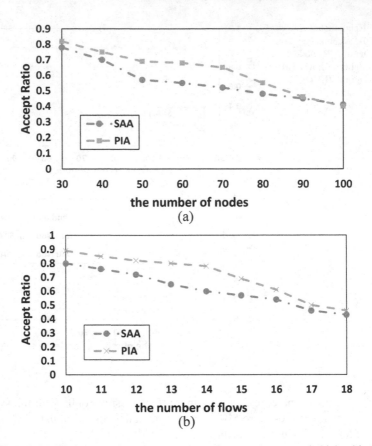

Fig. 5.10 Relationship between the acceptance rate and system parameters in high-critical mode. (**a**) Acceptance rate. (**b**) The number of nodes

conditions, we set $\aleph = 0.9 > \max\{\frac{c_i}{t_i}, i \in F\}$. If $\max\{\frac{c_i}{t_i}, i \in F\} \geq 0.9$, we will regeneration the system. Figure 5.10 illustrates (as Theorem 5.4 proves) that the acceptance rate in PIA is always better than in SAA.

Figure 5.11 shows the relationship between the system acceptance rate and system utilization. To study the relationship between these two elements, we increase system utilization by adjusting only the flow period. Subsequently, we obtain the same result as Fig. 5.10, which also verifies the correctness of Theorem 5.4.

5.6 Summary

In this chapter, we first introduce MRI nodes into mixed-criticality networks. Then, we analyze the transmission paths and obtain the candidate node set. Next, based on the characteristics of MRI nodes, we propose SAA and PIA to guarantee

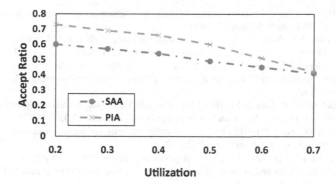

Fig. 5.11 Relationship between the number of MRI nodes and system utilization in high-critical mode

the network schedulability in low-criticality mode. By considering system cost, these two algorithms help to reduce the number of MRI nodes used. Finally, we analyze the schedulability of these two algorithms when the system switches to high-criticality mode. The simulation results show that our scheduling algorithms perform better than existing scheduling policies.

References

1. Lu CY, Saifullah A, Li B, Sha M, Gonzalez H, Gunatilaka D, Wu CJ, Nie LS, Chen YX (2016) Real-time wireless sensor-actuator networks for industrial cyber-physical systems. P IEEE 104(5):1013–1024
2. Lee EA, Seshia SA (2016) Introduction to embedded systems: a cyber-physical systems approach. MIT Press, Cambridge
3. Xiao HQ, Kong LS, Yuan CL, Xiao SP (2016) Stochastic optimization model and solution algorithm for blending procedure of process industrial. Inf Control 45(1):40–44
4. Saifullah A, Xu Y, Lu CY, Chen YX (2015) End-to-end communication delay analysis in industrial wireless networks. Trans Comput 64(5):1361–1374
5. Gesbert D, Shafi M, Shiu D, Smith PJ, Naguib A (2003) From theory to practice: an overview of MIMO space-time coded wireless systems. J Sel Area Commun 21(3):281–302
6. Hoydis J, Ten BS, Debbal M (2013) Massive MIMO in the UL/DL of cellular networks: how many antennas do we need? J Sel Area Commun 31(2):160–171
7. Lu L, Li G, Swindlehurst A, Ashikhmin A, Zhang R (2014) An overview of massive MIMO: benefits and challenges. J Sel Top Signal Process 8(5):742–758
8. Kong LH, Liu X (2015) mZig: enabling multi packet reception in zigBee. In: The annual international conference on mobile computing and networking. ACM, New York, pp 552–565
9. Hithnawi A, Li S, Shafagh H, Gross J, Duquennoy S (2016) Crosszig: combating cross-technology interference in low-power wireless networks. In: The international conference oninformation processing in sensor networks (IPSN). IEEE, Piscataway, pp 1–12
10. Cheng L, Gu Y, Niu JW, Zhu T, Liu C, Zhang Q, He T (2016) Taming collisions for delay reduction in low-duty-cycle wireless sensor networks. In: The IEEE international conference on computer communications. IEEE, Piscataway, pp 1–9

11. Yuan Y, He ZH, Chen M (2006) Virtual MIMO-based cross-layer design for wireless sensor networks. IEEE Trans Veh Technol 55(3):856–864
12. Zhao M, Yang YY, Wang C (2015) Mobile data gathering with load balanced clustering and dual data uploading in wireless sensor networks. IEEE Trans Mobile Comput 14(4):770–785
13. Liang W, Zhang XL, Xiao Y, Wang FQ, Zeng P, Yu HB (2011) Survey and experiments of WIA-PA specification of industrial wireless network. Wireless Commun Mobile Comput 11(8):1197–1212
14. Zu KK, DeLamare B, Haardt M (2014) Multi-branch Tomlinson-Harashima precoding design for MU-MIMO systems: theory and algorithms. Trans Commun 62(3):939–951
15. Jin X, Wang JT, Zeng P (2015) End-to-end delay analysis for mixed-criticality wirelesshart networks. IEEE/CAA J Autom Sin 2(3):282–289
16. Baruah S, Li HH, Stougie L (2010) Towards the design of certifiable mixed-criticality systems. In: Real-time and embedded technology and applications symposium (RTAS). IEEE, Piscataway, pp 13–22
17. Baruah S, Bonifaci V, D'Angelo G, Li HH, Marchetti-Spaccamela A, Megow N, Stougie L (2012) Scheduling real-time mixed-criticality jobs. Trans Comput 61(8):1140–1152
18. Ekberg P, Wang Y (2012) Bounding and shaping the demand of mixed-criticality sporadic tasks. In: Euromicro conference on real-time systems (ECRTS). IEEE, Piscataway, pp 135–144
19. Huang PC, Yang H, Thiele L (2014) On the scheduling of fault-tolerant mixed-criticality systems. In: The 51st annual design automation conference. ACM, New York, pp 1–6
20. Huang WH, Chen JJ (2016) Self-suspension real-time tasks under fixed-relative-deadline fixed-priority scheduling. In: The conference on design, automation & test in Europe, pp 1078–1083
21. Xia CQ, Jin X, Kong LH, Zeng P (2017) Bounding the demand of mixed-criticality industrial wireless sensor networks. IEEE Access 5:7505–7516
22. Saifullah A, Xu Y, Lu CY, Chen YX (2015) End-to-end communication delay analysis in industrial wireless networks. Trans Comput 64(5):1361–1374
23. Bini E, Buttazzo CC (2005) Measuring the performance of schedulability tests. Real-Time Syst 30(1):129–154

Chapter 6
Mixed-Criticality Scheduling on 5G New Radio

Abstract Compared to industrial wired networks, 5G can improve device mobility and reduce the cost of networking. However, the real-time performance and reliability of 5G NR (new radio) still need to be improved to satisfy industrial applications' requirements. In factories, the main factor that affects the performance of 5G NR is the unstable signal quality caused by high temperatures and metal. Although assigning dedicated resources to all transmissions and retransmissions is an effective method to improve the performance of 5G NR, the unstable signal quality causes the resources required for retransmissions to be uncertain. To address the problem, we introduce the mixed-criticality task model to 5G NR. When high-criticality packets cannot be transmitted, they are allowed to preempt the resources shared with low-criticality packets. The mixed-criticality scheduling problem of 5G NR is NP-hard. We formulate it as an OMT (optimization modulo theories) specification and propose a scheduling algorithm based on bin packing methods to make 5G NR satisfy industrial applications' requirements. Finally, we conduct extensive evaluations based on an industrial 5G testbed and random test cases. The evaluation results indicate that our algorithm makes communication reliability greater than 99.9% on unlicensed spectrum, and for most test cases, our algorithm is close to optimal solutions.

6.1 Background

Ultra reliable low latency communication (URLLC) is one of main application areas defined in 5G. It aims to provide low latency and ultra-high reliability for mission-criticality services, which widely exist in industrial systems. Since the performance of URLLC is comparable to some wired networks, industrial systems are adopting 5G in place of wired networks to improve the mobility of devices and reduce the cost of networking [1–3].

For URLLC, all transmissions and retransmissions of industrial data must be assigned the resources of 5G NR (new radio), in advance, by a scheduling algorithm that runs in the base station. The resources of 5G NR include time slots and frequency bandwidth. In factories, high temperatures, high humidity, and metal

© The Author(s) 2023
X. Jin et al., *Mixed-criticality Industrial Wireless Networks*, Wireless Networks,
https://doi.org/10.1007/978-981-19-8922-3_6

seriously affect signal qualities. Thus, more retransmissions are needed to guarantee the reliability of industrial communications. However, on the one hand, the signal quality is dynamic and unpredictable. Before data packets are actually transmitted, the scheduling algorithm cannot determine how many retransmissions and network resources are sufficient for the highly reliable communications. On the other hand, the scheduling algorithm cannot assign as many resources as possible to all packets because wireless network resources are limited. Therefore, in order to make 5G NR meet the real-time and reliability requirements of industrial systems, the scheduling problem of 5G NR needs to be studied.

5G NR supports two-dimensional (2D) time-frequency resources. Since the 2D resources are different from other systems, some researchers have begun to study the new scheduling problems for 5G. The work in [4] proves the NP hardness of the new problem and proposes an algorithm based on Lagrangian duality to guarantee the real-time performance of as many services as possible. The work in [5] aims at the same objective and proposes two heuristic algorithms to generate schedules quickly. The work in [6] applies machine learning to improve real-time performance and data rate. Similarly, the work in [7], based on machine learning, proposes an energy-efficient real-time scheduling algorithm. To make the proposed algorithms usable in actual systems, some real factors have been considered. The work in [8] considers the on-off operation of power amplifiers in the scheduling problem and proposes a sliding window-based algorithm to optimize the real-time performance and the energy efficiency for service transmissions. The work in [9] focuses on the impact of interference and channel estimation error on data rate. The work in [10, 11] studies how to guarantee the delay requirement under the minimum bandwidth. Although the scheduling problem of 5G NR has been studied more and more widely, the mixed-criticality scheduling problem of 5G NR has not been considered.

In industrial systems, control commands are the most important and must be delivered to devices in time, while some less important data, such as system logs and routine monitoring, can be delayed. Hence, under the limitation of 5G NR, the best way to schedule packets is to assign more resources to important packets, and allow the other packets to use the rest of the resources and the idle resources that have been assigned to the finished important packets. In other words, when the resources of 5G NR are insufficient, unimportant packets have to be discarded first. This process is typical of mixed-criticality scheduling. Some novel algorithms have been proposed to address the mixed-criticality scheduling problems of networks [12–17]. However, these algorithms cannot be used in the 5G NR model.

In this chapter, we introduce mixed criticality to 5G NR and propose a mixed-criticality scheduling algorithm to improve the real-time performance and reliability of industrial 5G networks. Much research has focused on mixed-criticality scheduling algorithms. However, the two main characteristics of our problem are not considered in other research.

1. On 5G NR, the available resources include time slots and frequency bandwidth, while in other mixed-criticality systems, the resources are time slots and processors. Compared to processors, frequency bandwidth is finer-grained and can be divided and converged. Although this characteristic contributes to the flexibility of scheduling algorithms, it also makes scheduling algorithms more complicated and more difficult to find optimized solutions.
2. In related work about mixed-criticality networks, e.g., [18, 19], when high-criticality packets are transmitting, all the low-criticality packets have to be discarded. However, our scheduling algorithm tries to guarantee the performance of all the high- and low-criticality packets. This difference makes us have no related work to refer to.

To solve the mixed-criticality scheduling problem of 5G NR, this chapter includes the following:

1. First, to rigorously state the problem, we propose a specification based on optimization modulo theories (OMT). The problem can be reduced to the bin packing problem. Therefore, our problem is at least NP-hard. Based on the specification, some off-the-shelf solvers can find optimal solutions.
2. Second, to improve the scalability of our work, we propose a heuristic, pseudolevel-packing algorithm to assign dedicated and shared resources to packets. In the algorithm, we extend real-time constraints and mixed criticality to the traditional bin packing problem, and analyze the sufficient condition and necessary condition for schedulability so that the solution space can be reduced effectively.
3. Third, we implement an industrial 5G testbed and evaluate our proposed algorithms. To compare all the algorithms under the same signal quality, we record signal qualities into trace files. Then, we conduct simulations based on the trace files and extensive test cases. The results indicate that our algorithm makes communication reliability greater than 99.9% on unlicensed spectrum, and for most test cases, our proposed algorithm is close to optimal solutions.

6.2 Problem Statement

The symbols used in this chapter are summarized in Table 6.1.

We consider the scheduling problem under one base station and many users. A data flow is from a user to the base station (or from the base station to a user). In the following, we ignore the direction of flows because flows in different directions have the same resource requirement. In the flow set F, all flows have the same period P. Each flow f_i generates a packet at time $j \times P$ ($j \in \mathbb{Z}$), and the packet must be delivered to its destination before its absolute deadline $(j + 1) \times P$. Since the packets contained in different periods have the same resource requirements, we only consider how to schedule packets in the first period P. After the first period, the subsequent schedules are periodically repeated.

Table 6.1 Symbols

Symbol	Definition
P	Period
F	Flow set
f_i	The i-th flow
χ_i	The highest criticality level of f_i
C_i	Transmission time durations
l_i	Frequency bandwidth
X	The highest criticality level of the network
F^e	Set of flows with $\chi_i = e$
τ_i	Packet generated by f_i
c_i^j	Transmission time duration of τ_i at criticality level j
x_i	Start time of τ_i
y_i	Start frequency of τ_i
Γ	Packet set
L	Bandwidth of the available resources
q_i	If τ_i is covered, $q_i = 1$.
ω^e	Weight of the e-th criticality level
Γ^e	Set of packets generated by the flows in F^e
$r_{i,j}$	If τ_i and τ_j cover or overlap each other, then $r_{i,j} = 1$.
Γ'	Ordered set of packets
τ_i'	The i-th packet in Γ'
l_i'	Frequency bandwidth of τ_i'
$c_i'^e$	Transmission time duration of τ_i' at criticality level e
h	Finish time of all placed packets
$a[y][x]$	If $a[y][x] = 1$, the corresponding resource is occupied
$\bar{y}$	The last row above row y
B^1, B^2, B^3	Three types of packets
c^1, c^2, c^3	Transmission durations of B^1, B^2 and B^3, respectively
l^1, l^2, l^3	Frequency bandwidths of B^1, B^2 and B^3, respectively
n^1, n^2, n^3	Number of local levels with lengths of c^1 (c^2 or c^3)
R	Resources actually occupied by all packets
E	Resources required by Algorithm 6.1
δ_j	Metric that indicates the difference between τ_j' and τ_i'
Γ_k'	Packet set returned by $Cover(k)$
N	Number of packets

Flow f_i is characterized by a three-tuple $< \chi_i, C_i, l_i >$, which denotes its highest criticality level, transmission time durations and frequency bandwidth, respectively. The highest criticality level of our network is set to X, and for each f_i, $1 \leq \chi_i \leq X$. We use F^e to denote the set of the flows with $\chi_i = e$. The packet generated by f_i is τ_i. $C_i = \{c_i^1, c_i^2, \ldots, c_i^{\chi_i}\}$ is the set of transmission time durations of τ_i. The flow with larger χ_i can occupy more time slots. At criticality level 1, τ_i is transmitted

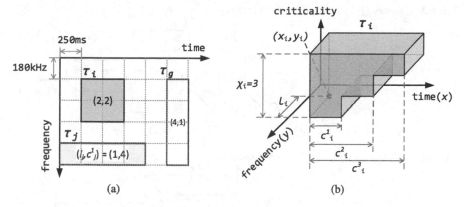

Fig. 6.1 Flow model. (a) (l_i, c_i^1) at criticality level 1. (b) τ_i at multiple criticality levels

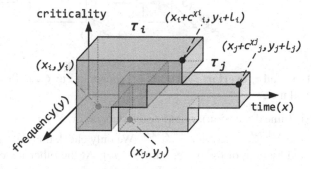

Fig. 6.2 Two packets overlap each other

once in one time slot, and at criticality level j, τ_i is transmitted j times in j time slots, i.e., $c_i^j = j \times c_i^1$. 5G numerology defines three subcarrier spacings (180, 360 and 720 kHz) and three corresponding slot lengths (1000, 500 and 250 μs). We define 15 kHz and 250 μs as the unit bandwidth and the unit slot length, respectively. Then, $(l_i, c_i^1) \in \{(1, 4), (2, 2), (4, 1)\}$ (as shown in Fig. 6.1a). We use x_i and y_i to denote the start time and start frequency of the transmission of τ_i, respectively. The illustration is shown in Fig. 6.1b. Initially, at the lowest criticality level, τ_i is transmitted once in time duration c_i^1. If τ_i is not sent successfully in the duration, then its critically level is set to 2. At criticality level 2, the transmission time duration is increased to c_i^2, and τ_i is transmitted again. Repeat this process until the transmission is successful. If the transmission still fails in the longest transmission time duration $c_i^{\chi_i}$, τ_i has to be discarded.

For any two packets τ_i and τ_j, the resources assigned to them are not allowed to *overlap* each other (Definition 6.1). An example is shown in Fig. 6.2. At criticality level 2, the two packets share resources. Then, τ_j may be preempted by τ_i. At the same criticality level, the packets are equally important and cannot preempt each other. Therefore, in feasible solutions, packet overlapping is not allowed.

Fig. 6.3 τ_i covers τ_j

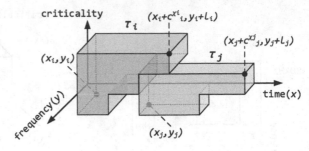

Fig. 6.4 Sufficient resources

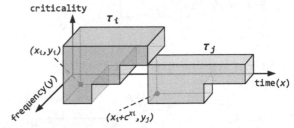

Definition 6.1 τ_j and τ_i **overlap** each other (as shown in Fig. 6.2), if the following conditions are all met.

1. On the x axis (time dimension), τ_i and τ_j overlap, i.e., $x_i \leq x_j \leq x_i + c_i^{\min\{\chi_i, \chi_j\}}$ or $x_i \leq x_j + c_j^{\min\{\chi_i, \chi_j\}} \leq x_i + c_i^{\min\{\chi_i, \chi_j\}}$. We only check the highest criticality level occupied by both of them, i.e., $\min\{\chi_i, \chi_j\}$. At the other lower criticality levels, they occupy fewer time slots. Even if they do not overlap at the lower criticality levels, they may overlap at criticality level $\min\{\chi_i, \chi_j\}$.
2. On the y axis (frequency dimension), τ_i and τ_j overlap, i.e., $y_i \leq y_j \leq y_i + l_i$, or $y_i \leq y_j + l_j \leq y_i + l_i$.

If τ_i *covers* τ_j (Definition 6.2), τ_j may not be sent. As shown in Fig. 6.3, when τ_i is being transmitted at criticality level 3, the resources assigned to τ_j are being occupied by τ_i. Then, τ_j has to be discarded.

Definition 6.2 τ_i **covers** τ_j (as shown in Fig. 6.3), if the following conditions are all met.

1. τ_i and τ_j do not overlap each other (Definition 6.1).
2. The highest criticality level of τ_i is larger than that of τ_j, i.e., $\chi_i > \chi_j$.
3. On the x axis (time dimension), τ_i and τ_j share the same time slots at the different criticality levels, i.e., $x_i \leq x_j \leq x_i + c_i^{\chi_i}$, or $x_i \leq x_j + c_j^{\chi_j} \leq x_i + c_i^{\chi_i}$.
4. On the y axis (frequency dimension), τ_i and τ_j share the same frequency bandwidth, i.e., $y_i \leq y_j \leq y_i + l_i$, or $y_i \leq y_j + l_j \leq y_i + l_i$.

Our objective is to send as many packets as possible. Once a packet is covered, it may be discarded. If there are sufficient resources, no packets can be covered. For example, x_j is set to $x_i + c_i^{\chi_i}$ (as shown in Fig. 6.4). Then, τ_i and τ_j do not cover or overlap each other. No matter which criticality level τ_i is at, τ_j is not affected. Only when resources are insufficient, τ_i is allowed to cover τ_j.

We formulate our problem as an OMT specification [20]. OMT is an extension of satisfiability modulo theories (SMT), which has been widely used to determine whether a specification is satisfiable or not. In addition to OMT supporting all the operators of SMT, it can also find an optimal objective. Our problem is to send as many packets as possible under scheduling constraints. Therefore, OMT is the best choice for our problem. The solution found by OMT not only satisfies scheduling constraints but also maximizes the number of packets sent. In the following specification, $\wedge$, $\vee$ and $\neg$ denote the logical operations of conjunction, disjunction and negation, respectively.

Each flow in F generates one packet. All the packets are included in the packet set Γ. These packets are transmitted in resources including a bandwidth of L and a time duration of P. The problem is how to determine (x_i, y_i) for each packet such that as many packets as possible are transmitted, and high-criticality packets are not covered as much as possible. Therefore, the objective is to minimize the weighted sum of the number of covered packets, as follows:

$$\min \sum_{\forall e \in [1, X]} (\omega^e \times \sum_{\forall \tau_i \in \Gamma^e} q_i), \tag{6.1}$$

where $q_i = 1$ indicates that τ_i is covered by a higher-criticality packet, ω^e is the weight of the e-th criticality level, and Γ^e includes the packets with $\chi_i = e$, i.e., $\Gamma^e = \{\tau_i | \forall f_i \in F, \chi_i = e\}$. In order to ensure that high-criticality packets are not covered as much as possible, we set that $\omega^1 = 1$ and $\forall e \in [2, X], \omega^e = \sum_{\forall g \in [1, e-1]} \omega^g \times |\Gamma^g| + 1$, i.e., ω^e is greater than the weighted sum of all lower-criticality packets. Hence, to minimize the objective, when there are not sufficient resources, low-criticality packets are covered first.

To check if τ_i and τ_j cover or overlap each other, we define the following function

$$Disjoint(i, c_i, j, c_j) = (x_i \geq x_j + c_j) \vee (y_i \geq y_j + l_j)$$
$$\vee (x_j \geq x_i + c_i) \vee (y_j \geq y_i + l_i). \tag{6.2}$$

The function only considers time-frequency resources, and criticality levels are reflected in c_i and c_j.

We use $r_{i,j}$ to bridge q_i and $Disjoint()$. $\forall f_i \in F, \forall f_j \in \{F^{\chi_i+1}, \dots, F^X\}$,

$$((r_{i,j} == 1) \wedge \neg Disjoint(i, c_i^{\chi_i}, j, c_j^{\chi_j}))$$
$$\vee ((r_{i,j} == 0) \wedge Disjoint(i, c_i^{\chi_i}, j, c_j^{\chi_j})), \tag{6.3}$$

$$1 \geq q_i \geq r_{i,j} \geq 0. \tag{6.4}$$

In Eq. (6.3), if τ_j and τ_i cover or overlap each other, $r_{i,j}$ is equal to 1. However, the following constraint 2) can avoid overlapping between two packets. Thus, only when τ_j covers τ_i, $r_{i,j} = 1$. Then, in Eq. (6.4), if there exists $r_{i,j} = 1$, then $q_i = 1$.

The minimizing problem has to respect the following constraints.

1. Range constraint: The ranges of variables used in the specification are as follows.

$$\forall f_i \in F, 0 \leq x_i < P - c_i^{\chi_i}, 0 \leq y_i < L - l_i,$$
$$q_i \in \{0, 1\}, r_{i,j} \in \{0, 1\}. \tag{6.5}$$

2. Overlap constraint: Any two packets are not allowed to overlap each other. Note that in Eq. (6.6) the transmission time duration of τ_j is at the highest criticality level of τ_i because the overlap of two packets only occurs at the same level.

$$\forall f_i \in F, \forall \tau_j \in \{\Gamma^e | \forall e \in [\chi_i, X]\}, Disjoint(i, c_i^{\chi_i}, j, c_j^{\chi_i}) = true. \tag{6.6}$$

A packet set is called *schedulable* if it has a feasible placement that meets all the constraints. When the objective is restricted to 0, the above problem is how to place all the packets into the rectangular area with dimensions $L \times P$. This is the same as the 2D bin packing problem, in which a set of rectangular items is packed into a 2D rectangular bin. The NP-hardness of the 2D bin packing problem has been proven [21]. Since our problem can be reduced to the 2D bin packing problem, it is at least NP-hard. Hence, there is no polynomial time algorithm for finding an optimal solution. Although the specification (Eq. (6.1)–(6.6)) can be solved by OMT solvers, e.g. Z3, for complicated systems, the execution time of solvers cannot be acceptable. Therefore, in the following section, we will propose a heuristic algorithm to schedule packets.

6.3 Scheduling Algorithm

Firstly, we introduce a basic scheduling algorithm that does not consider the time constraint P and does not support any packet being covered. Secondly, based on the basic scheduling algorithm, we analyze the sufficient condition and necessary condition for schedulability. Finally, we extend the basic algorithm based on the two conditions to support time constraints and packet covering.

6.3.1 Basic Scheduling Algorithm

The basic scheduling algorithm is a pseudolevel-packing algorithm (as shown in Algorithm 6.1). Fig. 6.5 shows an illustration. There are two types of packing

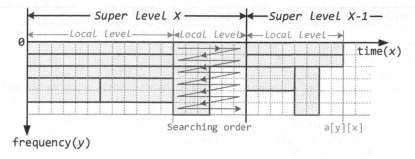

Fig. 6.5 Illustration of Algorithm 6.1

levels: *super level* and *local level*. A super level consists of multiple local levels. A local level is similar to the level that is widely used in multi-level bin packing algorithms, and its length is determined by the first packet placed at this local level. Algorithm 6.1 processes packets from the highest criticality level to the lowest criticality level (line 2). The packets with the same criticality level are placed at the same super level (lines 3–21). After all the packets in one criticality level are finished, a new super level is created to hold the packets in the next criticality level (lines 18–20). At each criticality level, the algorithm, first, sorts the packets according to the decreasing order of their transmission time durations (line 3), and then places packets in the same order (line 5). The single quotation mark on the symbols indicates that the symbols are sorted. For example, a sorted packet is denoted by τ_i', and its frequency bandwidth and transmission time duration are l_i' and $c_i'^e$, respectively. h is the finish time of all placed packets, and array $a[\ ][\ ]$ is used to indicate which resources are occupied. If $a[y][x] = 1$, the resource at time slot x and on frequency y is occupied; otherwise, the resource is available. At each local level, the algorithm searches resources first in the order of time slots and then in the order of frequency (lines 6–15). If an available resource is found, i.e., $a[y][x] = 0$, the upper left corner of τ_i' is placed at coordinates (x, y) (line 16). The algorithm does not need to check all the resources that will be occupied by the rectangle of τ_i' because according to the above searching order, the other resources must be available as long as $a[y][x]$ is available (Theorem 6.1). Then, based on the resources requested by τ_i', $a[\ ][\ ]$ is updated (line 17). If there is not sufficient resource in the current local level, a new local level is created (lines 7 and 8). Repeat this process until all packets are placed. Finally, the locations of all the packets and the finish time are returned to the calling function. The time complexity of the algorithm is $O(|\Gamma|Lh)$ because the algorithm traverses all the packets of Γ (lines 2 and 5) and all the resources of $L \times h$ (line 6). Since the basic scheduling does not support any packet being covered, the objective value is not considered in Algorithm 6.1.

Algorithm 6.1 Basic Scheduling Algorithm $BasicSch(\Gamma)$

Input: Γ
Output: $\forall(x_i, y_i)$ and h
1: $h = 0; a[\ ][\ ] = \{0\};$
2: **for each** $e = X$ to 1 **do**
3:    sort the packets of Γ^e according to the decreasing order of their c_i^e, where the first packet τ_1' has the largest $c_1'^e$;
4:    $x = h; y = 0; g = c_1'^e;$
5:    **for each** $i = 1$ to $|\Gamma^e|$ **do**
6:       **while** $(a[y][x] == 1)$ or $(y + l_i' \geq L)$ **do**
7:          **if** $(y + l_i' \geq L)$ **then**
8:             $h = x = h + g; y = 0; g = c_i'^e;$
9:          **else**
10:            $x = x + 1;$
11:            **if** $(x \geq h + g)$ **then**
12:               $x = h; y = y + 1;$
13:            **end if**
14:         **end if**
15:      **end while**
16:      $x_i' = x; y_i' = y;$
17:      $\forall j \in [y, y + l_i'), \forall k \in [x, x + c_i'^e), a[j][k] = 1;$
18:      **if** $(i == |\Gamma^e|)$ **then**
19:         $h = h + g;$
20:      **end if**
21:   **end for**
22: **end for**
23: **return** $\forall(x_i', y_i')$ and h;

Theorem 6.1 *In the process of placing τ_i', if $a[y][x] = 0$ and $y + l_i' < L$, then* $\forall j \in [0, l_i'), \forall k \in [0, c_i'^{X_i}), a[y + j][x + k] = 0.$

Proof Assuming that $\exists \hat{y} \in (y, y + l_i'), \exists \hat{x} \in (x, x + c_i'^{X_i}), a[\hat{y}][\hat{x}] = 1$. Recall that 5G numerology defines only three subcarrier spacings, 15 kHz, 30 kHz, and 60 kHz. Thus, at a super level, there are three types of packets. We use B^1, B^2 and B^3 to denote them, and their widths and lengths are $B^1 = (l^1, c^1)$, $B^2 = (l^2, c^2)$, and $B^3 = (l^3, c^3)$, respectively. Based on the definition of three subcarrier spacings, we know that $4 \times l^1 = 2 \times l^2 = l^3$ and $c^1 = 2 \times c^2 = 4 \times c^3$. Since $y + l_i' < L$, the frequency resource is sufficient. Hence, we do not need to discuss the frequency dimension. We use row $\bar{y}$ to denote the last row above row y and occupied by some packets. In the following, we discuss the three cases of row $\bar{y}$.

(1) **There is no row $\bar{y}$, and row y is the first row.** However, in Algorithm 6.1, the first row of a local level is fully occupied because only when the first packet is placed is a new local level with the same length as the packet created. Therefore, the unoccupied $a[y][x]$ is not at the first row, and row $\bar{y}$ must exist.

(2) **Row $\bar{y}$ is occupied by the same type of packets.** The same type of packets cannot be two B^1 because each local level contains at most one longest packet. If $a[y][x] = 0$ and $a[\hat{y}][\hat{x}] = 1$, then the placement is shown in Fig. 6.6a.

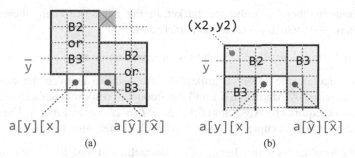

Fig. 6.6 Illustration of Theorem 6.1. (**a**) Illustration of Case (2). (**b**) Illustration of Case (3)

However, according to the searching order, the upper-left corner of the right packet should be at the point marked with a cross. Then, $a[\hat{y}][\hat{x}]$ is not occupied, i.e., $a[\hat{y}][\hat{x}] = 0$.

(3) **Row $\bar{y}$ is occupied by different types of packets.** The different types of packets must be B^2 and B^3. Since in Algorithm 6.1, the packets are sorted according to the decreasing order of their transmission time durations, the current packet τ_i' must be B^3. The illustration is shown in Fig. 6.6b. We know that $c^2 = 2 \times c^3$. Thus, two whole B^3 can be placed below B^2. There is no other packet with different transmission time durations in this problem. Therefore, the packet that is not aligned with B^2 does not exist, and $a[\hat{y}][\hat{x}]$ is not occupied, i.e., $a[\hat{y}][\hat{x}] = 0$.

To sum up, $a[\hat{y}][\hat{x}]$ is not consistent with its definition. The above assumption does not hold. Therefore, $\forall \hat{y} \in (y, y + l_i'), \forall \hat{x} \in (x, x + c_i'^{X_i}), a[\hat{y}][\hat{x}] = 0$.

$\square$

6.3.2 Analysis

We, first, analyze the networks with only one criticality level, and then extend to multiple criticality levels. To simplify the description, when only one criticality level is considered, we ignore the symbols about criticality levels. In the networks with only one criticality level, the lengths of local levels can be c^1, c^2 or c^3, and $L \geq l^3$. Assuming that in the result of Algorithm 6.1, there are n^1, n^2, and n^3 local levels with lengths of c^1, c^2, and c^3, respectively, and $n^1, n^2, n^3 \in \mathbb{N} \cap \{0\}$.

When only one criticality level is considered, Algorithm 6.1 has the following two properties.

Property 6.1 If $n^1 + n^2 + n^3 = 1$, and L approaches infinity, the resource utilization can be infinitesimal.

For example, there is only one packet in the network. Then, the resource utilization is $\frac{l_i \times c_i}{L \times c_i}$, which decreases as L increases.

Property 6.2 If $n^1 + n^2 + n^3 > 1$, the resource utilization must be greater than $\frac{1}{2}$.

Proof To calculate the resource utilization, we need to analyze the amount of resources required by Algorithm 6.1 and the amount of resources actually occupied. The resources required by Algorithm 6.1 is $E = L(n^1 c^1 + n^2 c^2 + n^3 c^3)$. The amount of resources actually occupied is analyzed in each of the following cases:

(1) For the local levels with a length of c^1, since the width of B^1 is the unit width, in the first $n^1 - 1$ local levels all the resources must be occupied. At the n^1-th local level, if $n^2 = n^3 = 0$, at least one B^1 is placed, i.e., only $l^1 \times c^1$ resources are occupied; if $n^2 \neq 0$, at most $(l^2 - 1) \times c^1$ resources are idle because $l^2 \times c^1$ resources are sufficient for the subsequent B^2; similarly, if $n^2 = 0$ and $n^3 \neq 0$, then $(l^3 - 1) \times c^1$ resources are idle. Thus, in the worst case, the amount of resources actually occupied in the n^1 local levels is

$$
R_1 = \begin{cases} Lc^1(n^1 - 1) + l^1 c^1, & \text{if } n^2 = n^3 = 0 \quad (R_{1.1}) \\ Lc^1 n^1 - (l^3 - 1)c^1, & \text{if } n^2 = 0, n^3 \neq 0 \quad (R_{1.2}) \\ Lc^1 n^1 - (l^2 - 1)c^1, & \text{others.} \quad (R_{1.3}) \end{cases}
$$

(2) For the local levels with a length of c^2, in the first $n^2 - 1$ local levels, at most $(L \bmod l^2)c^2(n^2 - 1)$ resources are idle. This is because after placing several B^2, the last $(L \bmod l^2)$ rows are not enough for the next B^2. At the n^2-th local level, if $n^3 = 0$, at least $l^2 c^2$ resources are occupied; otherwise, $\min\{(L - l^2), (l^3 - 1)\}c^2$ resources are idle because if L is very short, after placing B^2, the remaining resources may be less than $(l^3 - 1)c^2$. Thus, the amount of resources actually occupied in the n^2 local levels is

$$
R_2 = \begin{cases} Lc^2(n^2 - 1) - (L \bmod l^2) \times c^2(n^2 - 1) + l^2 c^2, & \text{if } n^3 = 0 \quad (R_{2.1}) \\ Lc^2(n^2 - 1) - (L \bmod l^2) \times c^2(n^2 - 1) + Lc^2 \\ \quad - \min\{(L - l^2), (l^3 - 1)\} \times c^2, & \text{others.} \quad (R_{2.2}) \end{cases}
$$

(3) For the local levels with a length of c^3, R_3 is similar to R_2, i.e.,

$$
R_3 = Lc^3(n^3 - 1) - (L \bmod l^3) \times c^3(n^3 - 1) + l^3 c^3.
$$

Thus, the total amount of resources actually occupied is $R = R_1 + R_2 + R_3$. Table 6.2 shows R and the lower bounds of utilization $\frac{R}{E}$ under different cases. The lowest bound in Table 6.2 is $\frac{1}{2}$. Therefore, the resource utilization must be greater than $\frac{1}{2}$.   □

Then, based on Properties 6.1 and 6.2, we analyze the sufficient condition for a packet set to be schedulable by Algorithm 6.1. Theorems 6.2 and 6.3 are about one

Table 6.2 Lower bound of resource utilization

n^1, n^2, n^3	$R = R_1 + R_2 + R_3$	Lower bound of $\frac{R}{E}$
0,0,-	R_3	$\frac{R}{E} > \frac{1}{2}$ (when $n^3 = 2, L = +\infty$)
0,-,0	$R_{2.1}$	$\frac{R}{E} > \frac{1}{2}$ (when $n^2 = 2, L = +\infty$)
0,-,-	$R_{2.2} + R_3$	$\frac{R}{E} > \frac{4}{7}$ (when $L = 7, n^2 = n^3 = 1$)
-,0,0	$R_{1.1}$	$\frac{R}{E} > \frac{1}{2}$ (when $n^1 = 2, L = +\infty$)
-,0,-	$R_{1.2} + R_3$	$\frac{R}{E} > \frac{4}{7}$ (when $L = 7, n^1 = n^3 = 1$)
-,-,0	$R_{1.3} + R_{2.1}$	$\frac{R}{E} > \frac{2}{3}$ (when $n^1 = n^2 = 1$)
-,-,-	$R_{1.3} + R_{2.2} + R_3$	$\frac{R}{E} > \frac{4}{7}$ (when $L = 7, n^1 = n^2 = 1, n^3 = +\infty$)

criticality level and multiple criticality levels, respectively. In addition, the necessary condition for a packet set to be schedulable is shown in Theorem 6.4.

Theorem 6.2 *For a packet set with only one criticality level, Algorithm 6.1 can find a feasible placement under the time constraint P, if* $\max\{\max_{\forall \tau_i \in \Gamma}\{c_i\}, \frac{2\sum_{\forall \tau_i \in \Gamma} l_i \times c_i}{L}\}$ $\leq P$.

Proof First, we discuss the placements with at least two local levels, i.e., $n^1 + n^2 + n^3 > 1$. In the proof of Property 6.2, R is the lower bound of resources actually occupied. Thus, the actual occupied resources $\sum_{\forall \tau_i \in \Gamma} l_i \times c_i$ is greater than or equal to R, i.e., $\sum_{\forall \tau_i \in \Gamma} l_i \times c_i \geq R > \frac{1}{2}E = \frac{1}{2}L(n^1 c^1 + n^2 c^2 + n^3 c^3)$. Rewriting, we obtain

$$\frac{2\sum_{\forall \tau_i \in \Gamma} l_i \times c_i}{L} > n^1 c^1 + n^2 c^2 + n^3 c^3.$$

If $\frac{2\sum_{\forall \tau_i \in \Gamma} l_i \times c_i}{L} \leq P$, then the actual length $n^1 c^1 + n^2 c^2 + n^3 c^3$ must be less than P, i.e., the placement is feasible under the time constraint P.

Then, from Property 6.1, we know that when $n^1 + n^2 + n^3 = 1$, $\frac{2\sum_{\forall \tau_i \in \Gamma} l_i \times c_i}{L}$ can be infinitesimal. However, the length of a local level is equal to the longest length of all the packets, i.e., $\max_{\forall \tau_i \in \Gamma}\{c_i\}$. Therefore, combining the above two cases, the sufficient condition is $\max\{\max_{\forall \tau_i \in \Gamma}\{c_i\}, \frac{2\sum_{\forall \tau_i \in \Gamma} l_i \times c_i}{L}\} \leq P$. $\square$

Theorem 6.3 (Sufficient Condition) *For a packet set with multiple criticality levels, Algorithm 6.1 can find a feasible placement under the time constraint P, if*

$$\sum_{\forall e \in [1, X]} (\max\{\max_{\forall \tau_i \in \Gamma^e}\{c_i^e\}, \frac{2\sum_{\forall \tau_i \in \Gamma^e} l_i \times c_i^e}{L}\}) \leq P. \tag{6.7}$$

Proof Since Algorithm 6.1 places super levels one by one, the total length is the sum of the lengths of all super levels. If the total length satisfies the time constraint P, the placement is feasible.   □

Theorem 6.4 (Necessary Condition) *For a packet set with multiple criticality levels, if Algorithm 6.1 find a feasible placement for a packet set under the time constraint P, then the packet set satisfies the following condition:*

$$\sum_{\forall \tau_i \in \Gamma} l_i \times c_i^{\chi_i} \leq L \times P. \tag{6.8}$$

Proof Note that in Algorithm 6.1 high-criticality packets are not allowed to cover low-criticality packets. Thus, if a packet set is schedulable, its resource utilization cannot be greater than 100%, i.e., $\frac{\sum_{\forall \tau_i \in \Gamma} l_i \times c_i^{\chi_i}}{L \times P} \leq 100\%$.   □

6.3.3   Improved Algorithm

In this subsection, we design our scheduling algorithm based on Theorems 6.3 and 6.4. The objective of our problem is to cover as few packets as possible, and low-criticality packets are covered first. Hence, we sort the packets of Γ according to the following rules: $\forall \tau_i, \tau_j \in \Gamma$,

- if $\chi_i > \chi_j$, then τ_i is before τ_j in the ordered set Γ';
- if $\chi_i = \chi_j$ and $c_i^{\chi_i} > c_j^{\chi_j}$, then τ_i is before τ_j in the ordered set Γ';
- if $\chi_i = \chi_j$, $c_i^{\chi_i} = c_j^{\chi_j}$ and $i < j$, then τ_i is before τ_j in the ordered set Γ'.

To optimize the objective, an effective method should cover the packets of Γ' from back to front. First, no packet can be covered, and the current packet set is checked whether it is schedulable or not under the time constraint P. If the packet set is not schedulable, the last packet $\tau_{|\Gamma'|}$ is covered. Then, if the new packet set is still unschedulable, the packets $\tau_{|\Gamma'|-1}$ and $\tau_{|\Gamma'|}$ are covered. Repeat this process until a schedulable packet set is found. However, this method may traverse all the packets. To improve the efficiency of our algorithm, we reduce the solution space based on Theorems 6.3 and 6.4, and adopt binary search instead of linear search. In Theorem 6.5, we proved that the solutions corresponding to the sorted packets are also ordered. Thus, the binary search can be used in our scheduling problem.

Our scheduling algorithm is shown in Algorithm 6.2. The variable k is used to denote that the packets after the k-th packet can be covered, and the function $Cover(k)$ (Algorithm 6.3) is to determine which packets are selected to cover these packets. In $Cover(k)$, for each packet τ_i' $(i > k)$, the metric δ_j indicates the area difference between τ_j' and τ_i' in the time-frequency coordinate (line 8). If τ_j' cannot fully cover τ_i', it is not an available selection (line 4). This is because if a small packet covers a big one, then the overflowing part of the big packet will change

---

Algorithm 6.2 Scheduling Algorithm with packet Covering (SAC)

Input: Γ', X, L, P
Output: $\forall (x_i, y_i)$
  1: $k = |\Gamma'|$;
  2: **do**
  3:     $\Gamma'_k = Cover(k)$; $k = k - 1$;
  4: **while** Γ'_k does not satisfy Eq. (6.8);
  5: $right = k + 1$;
  6: **do**
  7:     $\Gamma'_k = Cover(k)$; $k = k - 1$;
  8: **while** $k \geq 0$ and Γ'_k does not satisfy Eq. (6.7);
  9: $left = k + 1$; $middle = \lfloor \frac{left+right}{2} \rfloor$;
 10: **while** $left \neq middle$ **do**
 11:     $BasicSch(\Gamma'_{middle})$ returns $\forall (x_{middle,i}, y_{middle,i})$ and h;
 12:     **if** $h \leq P$ **then**
 13:         $left = middle$;
 14:     **else**
 15:         $right = middle$;
 16:     **end if**
 17:     $middle = \lfloor \frac{left+right}{2} \rfloor$;
 18: **end while**
 19: **if** $left = 0$ **then**
 20:     **return** FALSE;
 21: **end if**
 22: **return** $\forall (x_{left,i}, y_{left,i})$;

---

the shape of the small packet, and the schedulability will become worse. Then, to reduce the waste of resources, the packet with the minimal δ_j is selected to cover τ'_i (lines 14–15), and the covered packet τ'_i is marked in Γ' (line 16). If none of the packets have a valid δ_j, τ'_i cannot be covered even though it is allowed to be covered. Finally, $Cover(k)$ returns Γ' to Algorithm 6.2, called as Γ'_k.

Algorithm 6.2 traverses k from $|\Gamma'|$ to 1 (lines 1, 3 and 7). The first k that makes Γ'_k satisfy the necessary condition is the rightmost element of the binary search (lines 2–5), and then similarly, the first k that makes Γ'_k satisfy the sufficient condition is the leftmost element (lines 6–9). Then, in the binary search, for each middle element, $BasicSch(\Gamma'_{middle})$ is invoked to place the packet set and returns the finish time. If the finish time is not greater than time P, a better solution may be between $middle$ and $right$; otherwise, a feasible solution is between $left$ and $middle$ (lines 12–17). The leftmost element is always a feasible solution unless no packet set satisfies the sufficient condition. Thus, only when the leftmost element is 0, and the middle element does not search for any feasible solution, Γ cannot be scheduled (lines 19 and 20). Otherwise, the placement under the latest leftmost element is the final solution (line 22). In Algorithm 6.3, the number of iterations in lines 1, 2, 6 and 15 is $O(|\Gamma'|)$, $O(|\Gamma'|)$, $O(X)$ and $O(X)$, respectively. Thus, the time complexity of Algorithm 6.3 is $O(|\Gamma'|^2 X)$. In Algorithm 6.2, the number of invoking $Cover()$ is $O(|\Gamma'|)$ in the worst case. Hence, the time complexity of lines 1–9 is $O(|\Gamma'|^3 X)$. Then, in lines 10–18, the number of invoking $BasicSch()$

Algorithm 6.3 Covering Algorithm $Cover(k)$

Input: Γ', k
Output: Γ', and obj
1: **for** each $i = |\Gamma'|$ to k **do**
2:      **for** each $j = i - 1$ to 1 **do**
3:          **if** $\chi'_j \le \chi'_i$ or $l'_j < l'_i$ **then**
4:              $\delta_j = +\infty$;
5:          **else**
6:              **for** each $g = \chi'_i + 1$ to χ'_j **do**
7:                  **if** $c'^g_j \ge c'^{\chi'_i}_j + c'^{\chi'_i}_i$ **then**
8:                      $\delta_j = (c'^g_j - c'^{\chi'_i}_j) \times l'_j - c'^{\chi'_i}_i \times l'_i$;
9:                      break;
10:                 **end if**
11:             **end for**
12:         **end if**
13:     **end for**
14:     **if** $\exists \tau'_r, \delta_r \ne +\infty$ and $\nexists \tau'_m, \delta_m < \delta_r$ **then**
15:         $\forall g \in [1, \chi'_i], c'^g_r = c'^{\chi'_i}_r + c'^g_i$;
16:         τ'_i is marked as covered in Γ';
17:     **end if**
18: **end for**
19: calculate obj based on Eq. (6.1);
20: **return** Γ' and obj;

---

is $O(\log|\Gamma'|)$, and its time complexity is $O(\log|\Gamma'| \cdot |\Gamma'|Lh)$. Therefore, the time complexity of Algorithm 6.2 is $O(n^4)$.

Theorem 6.5 *If Γ'_k can be placed before time P, so can Γ'_{k-1}, and $Obj(\Gamma'_{k-1}) \ge Obj(\Gamma'_k)$; if Γ'_k cannot be placed before time P, neither can Γ'_{k+1}, and $Obj(\Gamma'_k) \ge Obj(\Gamma'_{k+1})$, where $Obj(\Gamma'_k)$ is the objective value of packet set Γ'_k calculated based on Eq. (6.1).*

Proof First, we prove that for our problem if Γ'_k has a feasible solution, then Γ'_{k-1} also has a feasible solution, and the objective value of Γ'_{k-1} is not greater than that of Γ'_k. The difference between Γ'_k and Γ'_{k-1} is that τ'_k may be covered in Γ'_{k-1}, while it cannot be covered in Γ'_k. There are two cases:

1. When τ'_k is covered in Γ'_{k-1}, we discuss all types of τ'_k as follows: if τ'_k is B^1, in the same local level, the places of the subsequent packets are moved from (x, y) to $(x, y - l^1)$; if τ'_k is B^2, in the same local level, the places of the subsequent packets are moved from (x, y) to $(x, y - l^2)$, $(x - c^2, y)$ or $(x + c^2, y - l^2)$; if τ'_k is B^3, in the same local level, the places of the subsequent packets are moved from (x, y) to $(x - c^3, y)$ or $(x + $ (the length of the local level) $- 1, y - l^3)$. In these cases, the subsequent packets are moved up and left in the same local level. Therefore, the finish time of Γ'_{k-1} is not later than that of Γ'_k. In addition, based on Eq. (6.1), if τ'_k is covered, $Obj(\Gamma'_{k-1}) = Obj(\Gamma'_k) + \omega^{\chi_k}$.

2. When τ'_k is not covered in Γ'_{k-1}, $\Gamma'_k = \Gamma'_{k-1}$, i.e., they have the same solution, and $Obj(\Gamma'_{k-1}) = Obj(\Gamma'_k)$.

Therefore, if Γ'_k can be placed before time P, so can Γ'_{k-1}, and $Obj(\Gamma'_{k-1}) \geq Obj(\Gamma'_k)$.

Second, for Γ'_k and Γ'_{k+1}, τ'_{k+1} can be covered in Γ'_k, but cannot be covered in Γ'_{k+1}. Packet τ'_{k+1} introduces more delay into the solution of Γ'_{k+1}. Hence, if the finish time of Γ'_k is greater than P, Γ'_{k+1} cannot finish before time P. For the objective value, if τ'_{k+1} is covered in Γ'_k, $Obj(\Gamma'_k) = Obj(\Gamma'_{k+1}) + \omega^{\chi'_{k+1}}$; otherwise, $Obj(\Gamma'_k) = Obj(\Gamma'_{k+1})$. Therefore, if Γ'_k cannot be placed before time P, neither can Γ'_{k+1}, and $Obj(\Gamma'_k) \geq Obj(\Gamma'_{k+1})$. $\qquad\square$

6.4 Performance Evaluations

In this section, we will evaluate our proposed algorithms based on a 5G testbed and extensive test cases. Five metrics are used in our evaluation: (1) **packet loss ratio (PLR)** is the ratio of the number of lost packets to the total number of sent packets; (2) **schedulable ratio** is the percentage of test cases for which an algorithm can find a feasible solution; (3) **objective value** is calculated based on Eq. (6.1); (4) **execution time** is the time required to find an optimized solution; and (5) **the number of calls to** $BasicSch(\ )$ reflects the effectiveness of Theorems 6.3 and 6.4.

Our proposed algorithm SAC is compared with the following methods:

1. **OMT** adopts the Microsoft solver Z3 [22] to solve our OMT specification (Eq. (6.1)–(6.6)). Although the solver can find the optimal solution, its execution time is unacceptable when the problem is complex. Thus, OMT only appears in simple test cases.
2. **T4** corresponds to the necessary condition of Theorem 6.4. We need an excellent baseline to illustrate the effectiveness of our algorithm. However, there is no method to find optimal solutions for complicated test cases. Therefore, when Z3 cannot find optimal solutions in an acceptable time, we replace OMT with T4. In T4, if a test case satisfies Theorem 6.4, it is considered schedulable. Hence, for schedulable ratios, T4 is better than optimal solutions when packets do not cover each other. For the objective of the scheduling problem, if in an ordered set Γ' the first k packets satisfy Theorem 6.4, we assume that the packets after these k packets are covered. Thus, the objective value of T4 is always better than optimal solutions. Note that T4 is an analysis-based method and cannot generate schedules. Hence, we do not consider its execution time and PLR.
3. **FFDH** (First-fit decreasing-height) [23] is a classical strip packing algorithm. In this chapter, FFDH sorts packets by order of non-increasing length, and then scans the local levels from left to right. Each packet is placed in the first level that has sufficient resources. FFDH does not support packet covering. Hence, we cannot calculate the original objective value of FFDH. The original objective

Table 6.3 Parameters

N	Number of packets
X	Number of criticality levels
L	Number of resources in the frequency dimension
P	Number of time slots in the time dimension

value is about the covered packets. In FFDH, this kind of packets includes those that cannot be placed before P. Therefore, when some packets cannot be placed before P, they are used to calculate the objective value based on Eq. (6.1).

4. **SACwoT3T4** is the same as SAC except that it does not adopt Theorems 6.3 and 6.4 to reduce the solution space.

All algorithms are written in C and run on a Windows workstation with a 3.7 GHz CPU and 64 GB memory. The parameters used in the evaluation are summarized in Table 6.3. In a test case, N packets are transmitted in a 3D space of dimensions $X \times L \times P$. For each packet, its criticality level is randomly selected in the range $[1, X]$, and the transmission time duration at the lowest criticality level and bandwidth are randomly selected in $\{(1, 4), (2, 2), (4, 1)\}$.

6.4.1 Evaluations Based on A Real Testbed

In this subsection, we evaluate packet loss ratios of different algorithms. The other metrics are not affected by the signal quality, and are more suitable for being tested through extensive test cases, which are shown in the next subsection. Our 5G testbed is shown in Fig. 6.7. It operates in the 2.4 GHz unlicensed ISM (industrial scientific and medical) band. The licensed spectrum is managed by mobile network operators, and developers and users cannot modify any strategy. Therefore, we adopt the unlicensed spectrum. If the strategies on the licensed spectrum are allowed to be customized, our proposed algorithms can be used without modification. Two 5G devices [24] are configured as a base station and a user, respectively. Since the signal quality is dynamic and unknown, to guarantee fairness for all algorithms, we trace the states of 7 subcarriers and 60,000 time slots, and then conduct trace-driven simulations.

First, in small networks, we compare SAC and OMT (as shown in Fig. 6.8). The parameter setting is $< N, X, L, P > = < 10, 4, 7, 20 >$, and CLx denotes the PLR at criticality level x. Since high-criticality packets can be transmitted more times, CL4 is the lowest. Both OMT and SAC successfully assign resources to the packets at criticality level 4. The average of CL4 in OMT and SAC are 0.072% and 0.067%, respectively. The slight difference between them is caused by the dynamics of signal quality. The average of CL1 in OMT and SAC are 16.7% and 17.0%, respectively. In SAC, since some low-criticality packets share resources with high-criticality packets, the CL1 of SAC is lower than that of OMT.

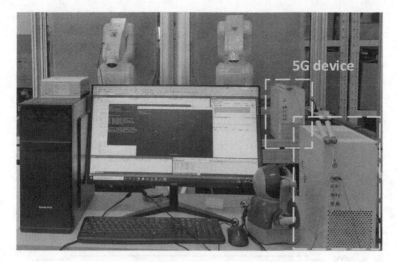

Fig. 6.7 5G testbed

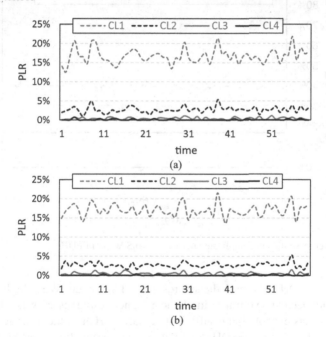

Fig. 6.8 Packet loss ratios in simple networks. (**a**) OMT. (**b**) SAC

Second, since OMT cannot solve complex problems in an acceptable time, we compare only SAC and FFDH in large networks (as shown in Fig. 6.9). The parameter setting is $< 80, 4, 7, 80 >$. Since the control period of the robotic arms in our testbed is 20 ms, and our 5G device supports 4 time slots in 1 ms, the parameter

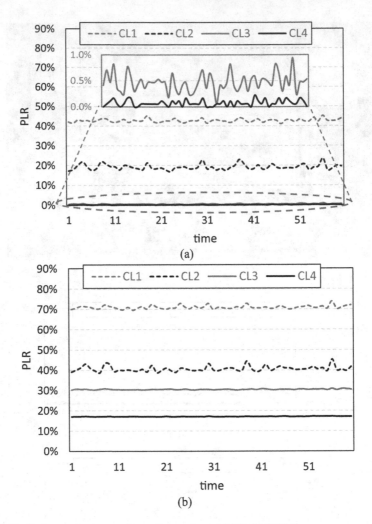

Fig. 6.9 Packet loss ratios in complicated networks. (**a**) SAC. (**b**) FFDH

P is set to 80. FFDH reserves dedicated resources for packets. When high-criticality packets do not need to be retransmitted, the assigned resources are wasted, and low-criticality packets cannot obtain sufficient resources. Hence, the average of CL1 is about 70.9%. In addition, FFDH does not consider criticality levels so that some high-criticality packets are discarded. Therefore, FFDH has higher PLRs even at the highest criticality level. The average of CL4 in FFDH is about 17.2%. SAC makes the best of resources to guarantee the requirements of packets. Hence, the average of CL4 in SAC is 0.071%, and for CL1, when the PLR of FFDH is 70.9%, SAC improves PLR to 42.7%. In SAC, the PLR of the highest criticality level is still about 0.07%, no matter which networks SAC is used in. Thus, SAC makes communication reliability greater than 99.9% under real-time constraints.

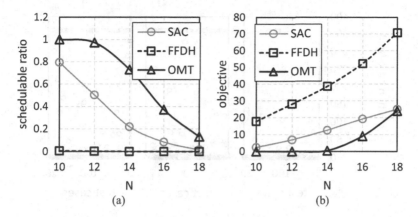

Fig. 6.10 Comparison with OMT. (**a**) Schedulable ratio. (**b**) Objective value

6.4.2 Evaluations Based on Random Test Cases

In the following, we will randomly generate extensive test cases to comprehensively evaluate the performance of our proposed algorithms. To make Z3 solvable, the parameter settings are $< [10, 18], 4, 7, 20 >$, and 12 processes run in parallel on our workstation. For each parameter setting, 200 test cases are randomly generated. The results are shown in Fig. 6.10. The objective value is the average of 200 test cases. As N increases, the schedulable ratios decrease, and the objective value increases because it is hard to find optimized solutions when more and more packets are transmitted on limited resources. The worst algorithm is FFDH. Since in FFDH, no packet can be covered, the limited resources are not sufficient to transmit all packets. OMT has the highest schedulable ratio and the lowest objective value. However, the execution time of OMT fluctuates greatly. For some simple test cases, Z3 cannot find solutions within 12 hours. Thus, we set the time limit of Z3 as 1 hour. The execution times are shown in Fig. 6.11. When $N < 16$, almost all test cases can be solved within the time limit; when $N \geq 16$, about 80% of test cases can be solved. Among all these test cases, the shortest execution time of OMT is 127 ms. Such a long execution time makes the network unable to respond to burst packets. However, when SAC is used to solve the same test cases, the longest execution time is only 0.19 ms. Even for the complicated test cases used in the following evaluations, the execution time of SAC is less than 3.7 ms, and the amount of memory space required is less than 1.2 MB. Therefore, SAC can quickly respond to burst packets and improve the flexibility of industrial networks.

Then, we increase the complexity of test cases. The parameter settings in Fig. 6.12 are $< [40, 100], 4, 10, 80 >$. For each parameter setting, 1000 test cases are randomly generated. When $N > 70$, SAC has a higher schedulable ratio than T4 because SAC allows packets to share resources. Although this causes some low-criticality packets to fail to be sent, compared to FFDH, SAC discards only

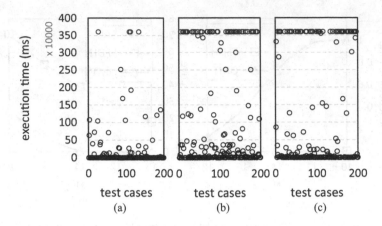

Fig. 6.11 Execution times of Z3. (**a**) $N = 14$. (**b**) $N = 16$. (**c**) $N = 18$

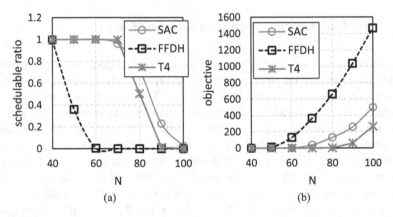

Fig. 6.12 Comparison under varying N. (**a**) Schedulable ratio. (**b**) Objective value

34% of packets, i.e., SAC makes a trade-off between schedulability and reliability. Therefore, in a network with limited resources, to schedule more high-criticality packets, SAC is the best choice.

In Fig. 6.13, we change X to illustrate its effect on schedulable ratios and objective values. The parameter settings are $< \{60, 70\}, \{3, 4, 5\}, 10, 80 >$. To show the results more clearly, the objective values of SAC under $N = 70$ and $X = 4$ are displayed in Fig. 6.14. Due to the limited resources, some packets have to be covered. Hence, the objective values of some test cases are greater than zero. In Fig. 6.13, as X increase, the schedulable ratio decreases, and the objective value increases. This is because more retransmissions lead to a more serious lack of resources. Furthermore, if a test case contains more packets than others, it will be more severely affected by X. For example, Fig. 6.13b and d ($N = 70$) has greater fluctuations than Fig. 6.13a and c ($N = 60$). Thus, if there are many packets in a

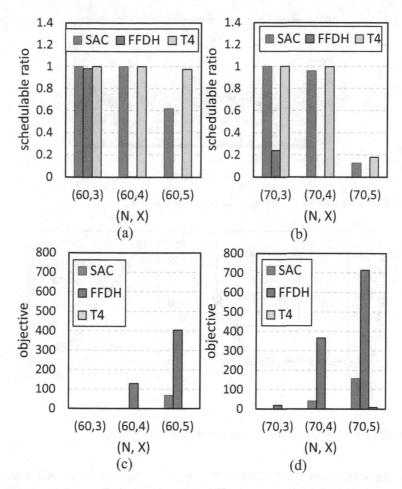

Fig. 6.13 Comparison under varying X. (**a**) Schedulable ratio. (**b**) Schedulable ratio. (**c**) Objective (**d**) Objective

network, X should be carefully determined based on the signal quality and should be as small as possible.

The parameter settings in Figs. 6.15 and 6.16 are $< 80, 4, [8, 16], 80 >$ and $< 80, 4, 10, [70, 95] >$, respectively. As L and P increase, the schedulable ratio increases, and the objective value decreases. This is because the more resources, the easier it is to schedule packets. In Figs. 6.15 and 6.16, when their total resources are the same, their schedulable ratios and objective values are similar. For example, in Fig. 6.15, when $L = 11$, the amount of resources is 880, and the schedulable ratio and objective value are 95% and 56, respectively. In Fig. 6.16, if the amount of

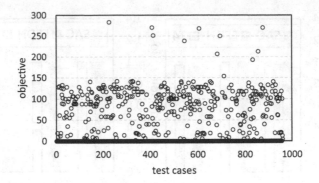

Fig. 6.14 The objective values of SAC under $N = 70$ and $X = 4$

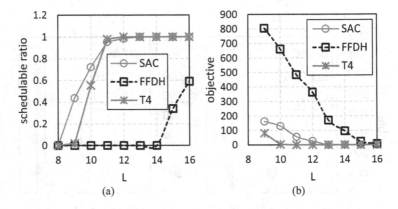

Fig. 6.15 Comparison under varying L. (**a**) Schedulable ratio. (**b**) Objective

resources is 880, then P is 88. When $P = 88$, the schedulable ratio and objective value are 94% and 58, respectively. Therefore, L and P have similar effects on schedulability and reliability, i.e., if the time constraint P of an industrial system cannot be relaxed, we can improve the system performance by increasing L.

In Fig. 6.17, we compare SAC and SACwoT3T4. SACwoT3T4 adopts binary search to find solutions in the whole solution space. Compared to SACwoT3T4, SAC reduces the solution space based on Theorems 6.3 and 6.4. Thus, SAC can significantly decrease the number of calls to function $BasicSch()$. When $N < 70$, SAC can directly find the optimal solution for almost all test cases. Therefore, Theorems 6.3 and 6.4 are effective.

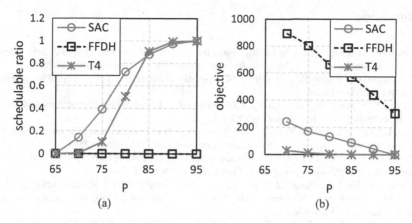

Fig. 6.16 Comparison under varying P. (**a**) Schedulable ratio. (**b**) Objective

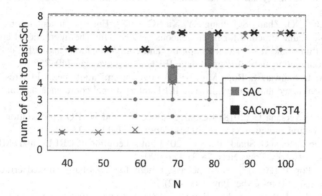

Fig. 6.17 Comparison between SAC and SACwoT3T4

6.5 Summary

This chapter focuses on the mixed-criticality scheduling problem of 5G NR. We present the mixed-criticality 5G NR model and formulate the problem as an OMT specification. Then, for the schedulability of the mixed-criticality scheduling problem, we analyze its sufficient condition and necessary condition. Based on the two conditions, we propose a scheduling algorithm. Finally, an industrial 5G testbed and extensive test cases are used to evaluate our proposed algorithm. The evaluation results indicate that our proposed algorithm can improve the real-time performance and reliability of 5G NR.

References

1. Huawei Technologies Co. Ltd. (2020) Unlocking the value of 5G industry applications. https://www.cio.com/article/3564155/unlocking-the-value-of-5g-industry-applications.html
2. 5G Industry Campus Europe (2020) 5G industry campus use cases. https://5g-industry-campus.com/use-cases/
3. Cheng JF, Chen WH, Tao F, Lin CL (2018) Industrial IoT in 5G environment towards smart manufacturing. J Ind Inf Integration 10:10–19
4. You L, Liao Q, Pappas N, Yuan D (2018) Resource optimization with flexible numerology and frame structure for heterogeneous services. Commun Lett 22(12):2579–2582
5. Nguyen TT, Ha VN, Le LB (2019) Wireless scheduling for heterogeneous services with mixed numerology in 5G wireless networks. Commun Lett 24(2):410–413
6. Zhang JX, Xu XD, Zhang KJ, Zhang BF, Tao XF, Zhang P (2019) Machine learning based flexible transmission time interval scheduling for eMBB and URLLC coexistence scenario. IEEE Access 7:65811–65820
7. Lei L, You L, He Q, Vu TX, Chatzinotas S, Yuan D, Ottersten B (2019) Learning-assisted optimization for energy-efficient scheduling in deadline-aware NOMA systems. Trans Netw 3(3):615–627
8. Sui WS, Chen XJ, Zhang SQ, Jiang ZY, Xu SG (2019) Energy-efficient resource allocation with flexible frame structure for heterogeneous services. In: The international conference on internet of things and IEEE green computing and communications and IEEE cyber, physical and social computing and IEEE smart data. IEEE, Piscataway, pp 749–755
9. Marijanovic L, Schwarz S, Rupp M (2019) A novel optimization method for resource allocation based on mixed numerology. In: The IEEE international conference on communications, pp 1–6
10. Anand A, De Veciana G (2018) Resource allocation and HARQ optimization for URLLC traffic in 5G wireless networks. J Sel Areas Commun 36(11):2411–2421
11. Anand A, De Veciana G, Shakkottai S (2020) Joint scheduling of URLLC and eMBB traffic in 5G wireless networks. Trans Netw 28(2):477–490
12. Gavrilut V, Pop P (2020) Traffic-type assignment for TSN-based mixed criticality cyber-physical systems. Trans Cyber Phys Syst 4(2):1–27
13. Gavrilut V, Zhao L, Raagaard ML, Pop P (2018) AVB-aware routing and scheduling of time-triggered traffic for TSN. IEEE Access 6:75229–75243
14. Harbin JR, Griffin DJ, Burns A, Bate IJ, Davis RI, Indrusiak LS (2018) Supporting critical modes in AirTight. In: Embedded and Real-Time Computing Systems and Applications. IEEE, Piscataway, pp 7–12
15. Burns A, Harbin J, Indrusiak L, Bate I, Davis R, Griffin D (2018) AirTight: a resilient wireless communication protocol for mixed criticality systems. In: The International Conference on Embedded and Real-Time Computing Systems and Applications. IEEE, pp 65–75
16. Xia C, Jin X, Kong L, Zeng P (2017) Bounding the demand of mixed criticality industrial wireless sensor networks. IEEE Access 5:7505–7516
17. Jin X, Xia C, Xu H, Wang J, Zeng P (2016) Mixed criticality scheduling for industrial wireless sensor networks. Sensors 16(9):1376–1396
18. Novak A, Sucha P, Hanzalek Z (2016) Efficient algorithm for jitter minimization in time-triggered periodic mixed-criticality message scheduling problem. In: The international conference on real-time networks and systems, pp 23–31
19. Novak A, Sucha P, Hanzalek Z (2019) Scheduling with uncertain processing times in mixed-criticality systems. Eur J Oper Res 279(3):687–703
20. Kovasznai G, Biro C, Erdelyi B (2017) Generating optimal scheduling for wireless sensor networks by using optimization modulo theories solvers. In: The international workshop on satisfiability modulo theories, pp 15–27
21. Lodi A, Martello S, Monaci M (2002) Two-dimensional packing problems: a survey. Eur J Oper Res 141(2):241–252

22. Bjorner N, Phan AD (2014) vZ - Maximal Satisfaction with Z3. In: The 6th international symposium on symbolic computation in software science. pp 1–9
23. Coffman J, Edward G, Garey MR, Johnson DS, Tarjan RE (1980) Performance bounds for level-oriented two-dimensional packing algorithms. SIAM J Comput 9(4):808–826
24. Yunzhi Ruantong Ltd. (2021) 5G URLLC device user guide. http://yunzhiruantong.com/?m=home&c=View&a=index&aid=176

Chapter 7
Conclusions and Future Directions

Abstract In this chapter, we summarize the book and provide three potential future directions for mixed-criticality industrial wireless networks.

7.1 Conclusions

In this book, we have introduced mixed criticality into industrial wireless networks, and presented analysis methods and scheduling algorithms to improve the QoS of industrial wireless networks.

In Chap. 1, we presented the concept of mixed-criticality industrial wireless networks. Industrial wireless networks have to meet the stringent QoS requirements of industrial applications. However, due to the openness of the wireless environment, the available network resources are limited. Mixed criticality is an advanced theory that makes limited resources fully utilized and can help industrial wireless networks improve their QoS.

In Chap. 2, we presented an end-to-end delay analysis method for fixed priority scheduling in mixed-criticality WirelessHART networks, which can be used to determine whether all flows can be delivered to destinations within their deadlines. In evaluations, we compared our analysis results with simulations and a testbed. The results show that the pessimism of our analysis is acceptable and reliable.

In Chap. 3, we focused on the analysis method under the EDF policy. Firstly, we proposed a novel network model that can switch routing strategies based on the criticality of networks. When errors or accidents occur, the network switches to high-criticality mode and low-level critical tasks are abandoned. Secondly, we analyzed the demand bound of mixed-criticality industrial wireless networks under the EDF policy and formulated network demand bounds in each criticality mode. Thirdly, we tightened the demand bound by analyzing carry-over jobs and classifying the number of conflicts to improve analysis accuracy. Simulation results demonstrate that the presented methods can estimate the schedulability efficiently.

Mixed-criticality data flows coexist in advanced industrial applications. They share the network resource, but their requirements for the real-time performance and reliability are different. In Chap. 4, we proposed a scheduling algorithm to guarantee

X. Jin et al., *Mixed-criticality Industrial Wireless Networks*, Wireless Networks,
https://doi.org/10.1007/978-981-19-8922-3_7

their different requirements, and then analyzed the schedulability for this scheduling algorithm. Simulation results show that our scheduling algorithm and analysis have more performance than existing ones.

In Chap. 5, we first introduced MRI nodes into mixed-criticality networks. Then, we analyzed the transmission paths and obtained the candidate node set. Next, based on the characteristics of MRI nodes, we proposed the algorithm SAA and the algorithm PIA to guarantee the network schedulability in low-criticality mode. By considering system cost, these two algorithms help to reduce the number of MRI nodes used. Finally, we analyzed the schedulability of these two algorithms when the system switches to high criticality mode. The simulation results show that our scheduling algorithms and analysis perform better than the existing scheduling policy.

In Chap. 6, we focused on the mixed-criticality scheduling problem of 5G NR. We presented the mixed-criticality 5G NR model and formulated the problem as an OMT specification. Then, for the schedulability of the mixed-criticality scheduling problem, we analyzed its sufficient condition and necessary condition. Based on the two conditions, we proposed a scheduling algorithm. Finally, an industrial 5G testbed and extensive test cases were used to evaluate our proposed algorithm. The evaluation results indicate that our proposed algorithm can improve the real-time performance and reliability of 5G NR.

7.2   Future Directions

There are many potential future directions for mixed-criticality industrial wireless networks. Here, we list three promising directions, as follows.

- *Criticality identifier for industrial communications.* In mixed-criticality industrial wireless networks, high-criticality communications are assigned more network resources. However, which communications should be high-criticality? Criticality is not an inherent property of industrial communications and must be identified according to some rules. The identification rules determine whether there is a good match between industrial requirements and the goals of algorithms. If criticality levels are not correctly identified, even the optimal scheduling algorithm cannot meet industrial requirements.
- *Efficient scheduling algorithm when the criticality level is switched.* On the one hand, the system state space is extremely complex, and there is no way to enumerate all the switching opportunities. On the other hand, when the criticality level is switched, there is no time to invoke a scheduling algorithm again. Under these restrictions, when the criticality level is switched, only the simple scheduling rule can be applied, such as preempting resources from the nearest low-criticality transmissions (as shown in Chaps. 4, 5 and 6). There is still a gap between these simple rules and optimal solutions. Therefore, efficient and effective scheduling algorithms should be studied.

- *Testbed supporting mixed criticality.* Currently, mixed-criticality studies are evaluated based on simulations or real testbeds. Although the simulation is more flexible, it is too ideal to fully demonstrate the situation in real-world scenarios. A testbed can provide more comprehensive evaluations. However, there are no mixed-criticality hardware platforms and software protocol stacks. An easy-to-use testbed will facilitate research and development.

Printed in the United States
by Baker & Taylor Publisher Services